Studies in Computational Intelligence

Volume 964

The series "Studies in Computational Intelligence" (SCI) publishes new developments and advances in the various areas of computational intelligence—quickly and with a high quality. The intent is to cover the theory, applications, and design methods of computational intelligence, as embedded in the fields of engineering, computer science, physics and life sciences, as well as the methodologies behind them. The series contains monographs, lecture notes and edited volumes in computational intelligence spanning the areas of neural networks, connectionist systems, genetic algorithms, evolutionary computation, artificial intelligence, cellular automata, self-organizing systems, soft computing, fuzzy systems, and hybrid intelligent systems. Of particular value to both the contributors and the readership are the short publication timeframe and the world-wide distribution, which enable both wide and rapid dissemination of research output.

Indexed by SCOPUS, DBLP, WTI Frankfurt eG, zbMATH, SCImago.

All books published in the series are submitted for consideration in Web of Science.

More information about this series at http://www.springer.com/series/7092

Tom Rutkowski

Explainable Artificial Intelligence Based on Neuro-Fuzzy Modeling with Applications in Finance

Springer

Tom Rutkowski
Vandenroot Institute
Jersey City, NJ, USA

ISSN 1860-949X ISSN 1860-9503 (electronic)
Studies in Computational Intelligence
ISBN 978-3-030-75520-1 ISBN 978-3-030-75521-8 (eBook)
https://doi.org/10.1007/978-3-030-75521-8

This Springer imprint is published by the registered company Springer Nature Switzerland AG
The registered company address is: Gewerbestrasse 11, 6330 Cham, Switzerland

Preface

In recent years, the research community of artificial intelligence has focused primarily on increasing the performance of algorithms. Models based on deep neural networks have shown spectacularly good results, especially in image recognition, automatic text translation, and voice recognition. The use of Generative Adversarial Networks enabled the automatic generation of images and music. Thanks to the development of these techniques, we can observe enormous progress in medical diagnostics or autonomous cars. The problem is that these models work on the principle of "black boxes". This means that while they perform very well, it is very difficult to justify exactly how they work. There are many cases where the lack of explanation is unacceptable because of the consequences of making the wrong decision. This may include medicine, law, finance, or the control of autonomous cars. In high-stakes, decision-making processes, people should require explanations with every recommendation or prediction regardless of the source of a recommendation—a human or a machine.

This book focuses on proposing techniques that will make recommendation systems both accurate and explainable with the emphasis on the financial sector because it is in particular very challenging from the perspective of data. Objectives are also different in Finance in comparison to general technological problems. The most popular tech problems are related to Image Recognition, Natural Language Processing, Computational Advertising, or Robotics. It is very different from solving specific problems in Finance regarding Forecasting, Valuation, Trading and Asset Management, or Decision-Making in Banking. Usually, in Finance, data are non-stationary, small or medium-size, and typically quite noisy. The uncertainty is high, as are the stakes, therefore interpretability of results is desired by end-users. Not to mention that explainability is more and more often required by regulators. Because of these reasons, it is essential to develop methods that can be used to solve problems with the above objectives and constraints. The author of this book believes and proves that using techniques based on the Neuro-Fuzzy approach is the way to go in such challenging cases.

Jersey City, USA | Tom Rutkowski

Acknowledgments

I wish to express my sincere appreciation to Dr. Radosław Nielek, who noticed, from the very beginning, the importance of my research and had enough patience to guide me over the years. His experience in incorporating social aspects to computer science was inspirational and changed my perspective.

I would like to thank my wife, Joanna, for a daily dose of motivation and encouraging me to keep going. I am glad that we could dedicate a big part of our lives for research in different fields and learn from each other during the process.

I am grateful to my parents—Prof. Danuta Rutkowska and Prof. Leszek Rutkowski, who always support me. I could not be happier that I had an opportunity to work and publish together. I admire them. All their experience and challenging conversations made my ideas and research much more advanced.

I wish to pay my special regards to Mark Zurada, my dear friend and business partner, for trusting me that it is possible to commercialize my research. Mark's perspective and input pushed me to make my research matter not only for other researchers but also for end-users of products based on Explainable Recommender Systems.

I would like to recognize the invaluable assistance that Dr. Krystian Łapa and Dr. Maciej Jaworski provided during my research. I would also like to show my gratitude to Prof. Robert Nowicki for suggesting alternative methods to developing Explainable AI.

I am grateful to Prof. Janusz Kacprzyk and Prof. Ewaryst Rafajłowicz for their valuable comments on the first version of the manuscript.

Last but not least, I take this opportunity to show my gratitude to Dr. hab. Jerzy Paweł Nowacki, professor, and rector of Polish-Japanese Academy of Information Technology for the opportunity to cooperate in a truly inspirational environment. Thanks to the support from PJAIT, I had a chance to present my research at conferences and discuss my results with the scientific community around the world.

Contents

List of Figures

List of Tables

Chapter 1
Introduction

1.1 The Purpose of This Book

The goal of the research presented in this book was to propose and develop explainable algorithms applicable to recommender systems.

It will be shown that:

1. The use of a Neuro-Fuzzy approach in content-based recommender systems provides a possibility to generate understandable explanations for each recommendation.
2. Recommender systems based on a Neuro-Fuzzy approach can be accurate, interpretable, and transparent.
3. Fuzzy modeling can be effectively applied to stock market recommendations.

This book presents new content-based recommendation systems developed by implementing a neuro-fuzzy approach (see e.g. [73, 113]) leading to interpretable recommenders. By adequately performing pre-processing of nominal data (e.g. movie genres) and transforming them into values of linguistic variables, we can construct a neuro-fuzzy system. The initial fuzzy rules are generating using the Wang-Mendel method [119], which divides the input space into fuzzy regions and then applies a table-lookup scheme to extract interpretable rules. The resulting system can be fine-tuned, using the backpropagation or one of the population-based methods. To our best knowledge, the method presented in this book is the first successful implementation of neuro-fuzzy techniques in content-based recommender systems.

Although this book is concerned with specific applications dealing with the MovieLens and asset management companies data (Chaps. 3 and 4, respectively), it has a great potential to solve other problems, in particular when interpretability is a critical matter.

T. Rutkowski, *Explainable Artificial Intelligence Based on Neuro-Fuzzy Modeling with Applications in Finance*, Studies in Computational Intelligence 964,
https://doi.org/10.1007/978-3-030-75521-8_1

As a matter of fact, the results presented in Chap. 4 were applied to the commercial product created by the author of this book and exclusively distributed by Senfino AI LLC in New Jersey.[1]

1.2 The Pursuit of Explainable Artificial Intelligence

Artificial Intelligence (AI) and Machine Learning (ML) research have been booming over the last few years since Deep Learning applications were introduced. As a Deep Learning approach is based on neural networks, it is very difficult to explain its behavior. The future of artificial intelligence cannot ignore the need for interpretable, transparent, and explainable models. It is simply not possible to rely on models that are mostly black boxes as seen in Fig. 1.1. People demand explanations from other people, but for some reason, they are fine dealing with algorithms that do not provide any kind of explanation in addition to predictions or recommendations. Receiving end of a recommendation should require explanation regardless of the source of a recommendation—a human or a machine. Fully explainable models should take into consideration: bias, fairness, transparency, safety, causality, engineering.

In 2016, Defense Advanced Research Projects Agency (DARPA)[2] published a description of their Explainable Artificial Intelligence Program (see [42]). Resources and information about this program are available at www.darpa.mil/program/explainable-artificial-intelligence, by Matt Turek. David Gunning, through this initiative, introduced and popularized the term XAI in the research community and the industry. The program influenced many researchers, including the research and experiments described in this book. DARPA distinguished two main areas of research—research on Explainable Models and research on the Explanation Interface—see Fig. 1.2.

There are few crucial terms related to Explainable Artificial Intelligence:

- interpretability (Interpretable ML)—*being able to infer how the system works*,
- understandability—*being able to understand a prediction or recommendation—recommendations should be different depending on the level of expertise of the user (doctor versus patient)*,
- responsibility (Responsible AI)—*who should feel responsible for using the outcome of a model—a user? a creator? a regulator?*,
- accountability (Accountable AI)—*who should be accountable for negative consequences of using the model? How to decide who should be accountable?*,
- transparency (Transparent AI)—*an opposite to black-box models—every component of the system is transparent*,
- fairness (Fair AI)—*a system is not biased and treats all its users in the same way*,

[1] www.senfino.ai.

[2] https://www.darpa.mil/.

Fig. 1.1 *Source* (New Yorker, December 30, 2015)

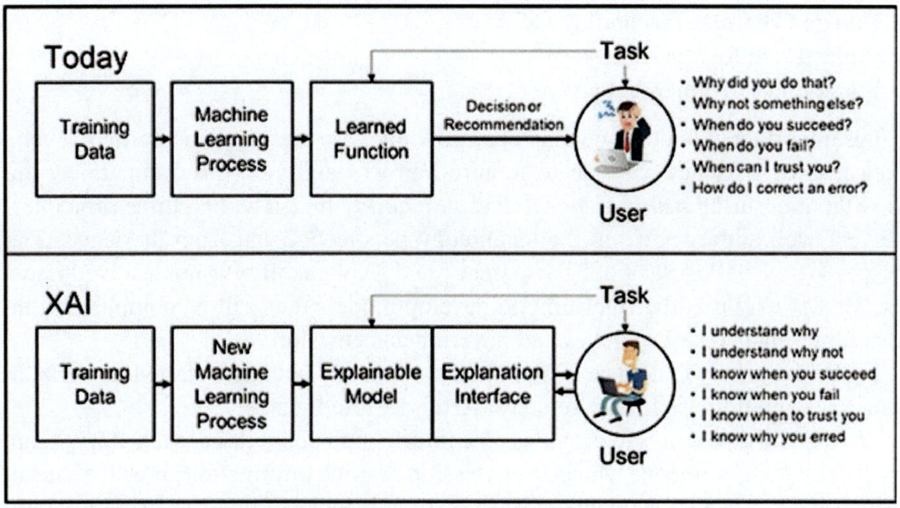

Fig. 1.2 XAI Concept (DARPA XAI program, 2016)

- ethics (Ethical AI)—*how to ensure that all parts of the systems and people involved behave ethical*,
- safety (Safe AI)—*a system knows when the consequences of its recommendations or predictions can be unpredictable so it would not proceed if it is to risky*,
- causality—*system that can show cause-effect relationships and act on them, not only based on correlations*,
- simulatability—*a human is able to simulate and reason about the system's entire decision-making process* [75],
- trust—*a system which can explain itself, and is right more often than not, can gain trust from a user*,
- Human-in-the-loop AI—*a system that is completing a user, not replacing her.*

Researchers from Berkman Klein Center for Internet and Society at Harvard University[3] created a map of many different frameworks regarding Explainable AI, Transparent AI, regulations and put them under one umbrella that they called "Principled Artificial Intelligence". That is the most recent attempt to organize issues and terminology as it is shown in Fig. 1.3. Authors of the paper [38] distinguished 8 principles:

1. Privacy
2. Accountability
3. Safety and Security
4. **Transparency and Explainability**
5. Fairness and Non-discrimination
6. Human Control of Technology
7. Professional Responsibility
8. Promotion of Human Values.

It is important to understand that those principles are presented in a form of layers for a reason. It is not possible to require Fairness and Non-discrimination from AI without ensuring *transparency and explainability* first. The first three principles are very well addressed from the technology perspective and from the legislation perspective as well. A lack of Transparency and Explainability is not a solved issue yet. That is why it is the right time to develop models that will be compliant with what representatives of societies and governments envision.

Figure 1.4 shows that since 2016 regulators in many countries started to work on guidelines and policies for the future of Artificial Intelligence.

"To explain an event is to provide some information about its causal history. In an act of explaining, someone who is in possession of some information about the causal history of some event—explanatory information, I shall call it—tries to convey it to someone else." According to this definition [59], explanation requires some degree of causality. After all, it is answering a "why" or "how" question. Is it possible to justify a recommendation or prediction if a model is not interpretable? How is it possible to explain its behaviour if a model is a black box?

[3] https://cyber.harvard.edu/.

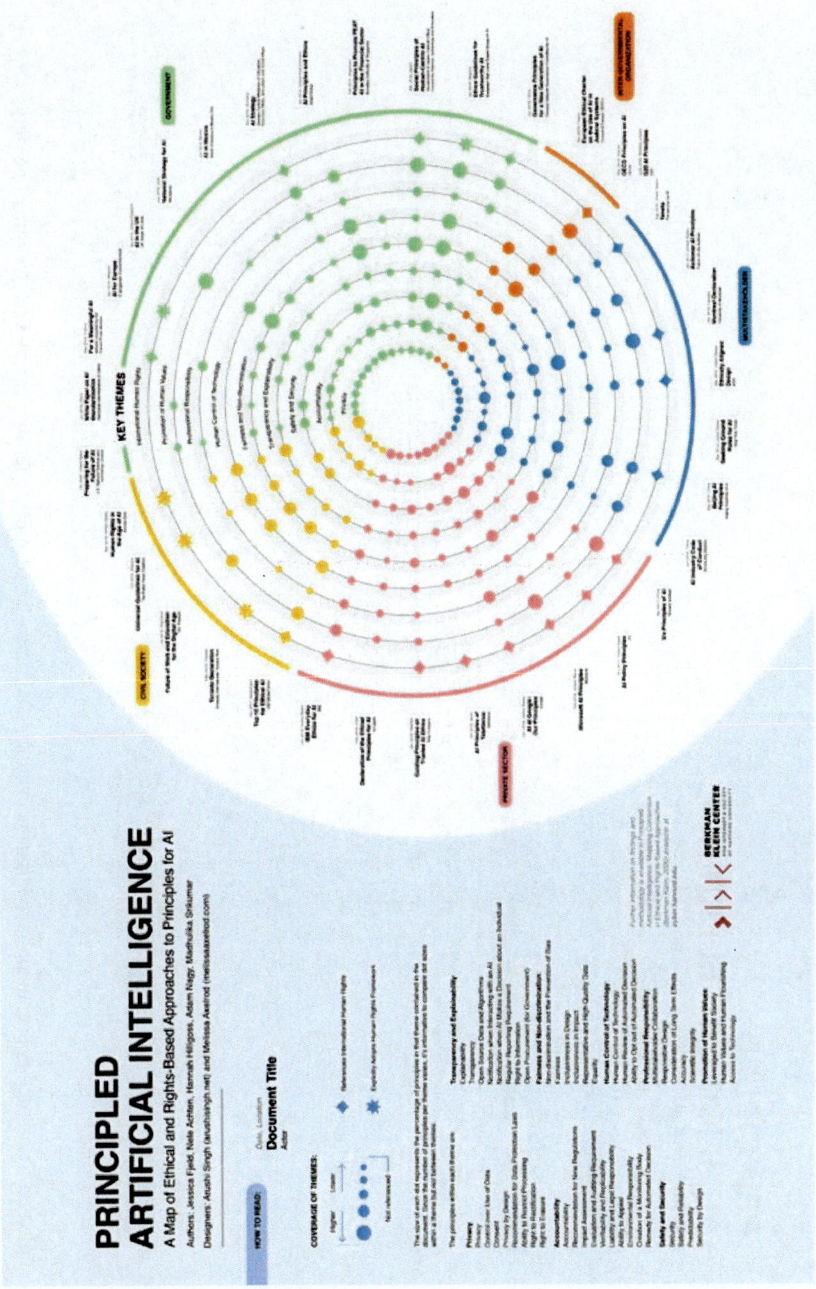

Fig. 1.3 Principled artificial intelligence: a map of ethical and rights-based approached to principles for AI (https://cyber.harvard.edu/publication/2020/principled-ai)

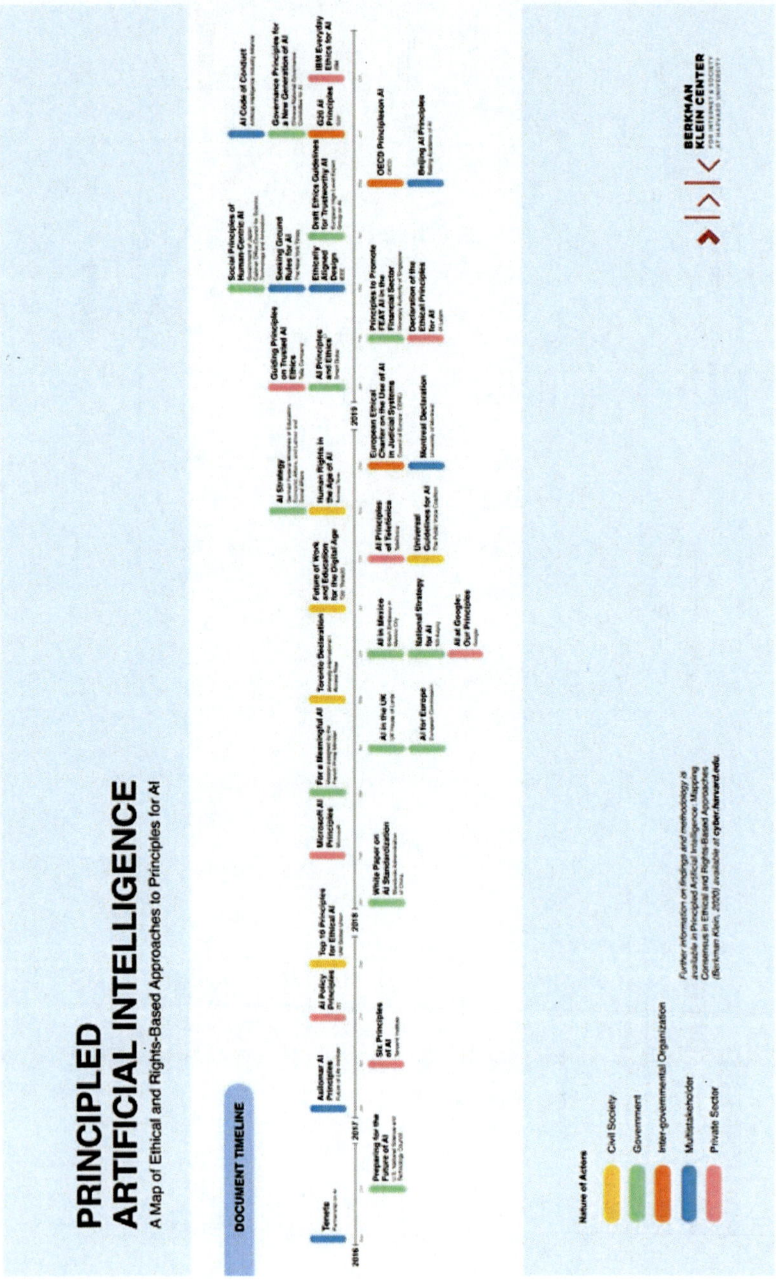

Fig. 1.4 Principled artificial intelligence: mapping consensus in ethical and rights-based approaches to principles for AI—timeline (https://cyber.harvard.edu/publication/2020/principled-ai)

There are several techniques under "explainer" or "explanator" umbrella, such as LIME (Local Interpretable Machine Explanations) [97] or SHAP (SHapley Additive exPlanations) [94]. There are many opinions about explainability. Some experts, even Geoffrey Hinton, the inventor of Deep Learning, said that the best AI systems are so complex that we should not expect them to be explainable and we should learn how to trust them without that. On the other hand, his partner, Yoshua Bengio, said that Deep Learning is not the ultimate machine learning solution and we should work on causality and explainability.

Because of the hype that Deep Learning brought to the research community, many researchers work on solutions for explaining deep learning. But explaining how we think the model came up with a prediction or recommendation is not explaining the model itself. All those techniques are nothing else than approximators working on top of black box models. Analyzing many different pairs of inputs and outputs can indeed lead to some conclusions about how the model behaves but it does not mean that the underlying model is all of a sudden transparent. It could be even more damaging in many business scenarios to think that explainers make black boxes transparent.

We should assume, that under some circumstances, using an unfortunate set of inputs, model can behave unpredictably. Therefore, the author of this book has a strong reservation to use so called model-agnostic explainers. What is the alternative? Is it possible to build interpretable, explainable model that is accurate enough to use in practice? Cynthia Rudin, in her positioning paper "Stop Explaining Black Box Machine Learning Models for High Stakes Decisions and Use Interpretable" [99] argues that this is exactly what the research community should be working on.

New regulations, like General Data Protection Regulation (GDPR) in European Union or California Consumer Privacy Act (CCPA) are also responsible for accelerating research on Explainable AI.

> Article 22 EU GDPR "Automated individual decision-making, including profiling"
>
> *"The data subject shall have the right not to be subject to a decision based solely on automated processing, including profiling, which produces legal effects concerning him or her or similarly significantly affects him or her."*

Similar concerns have been expressed by various researchers. Zachary Lipton made an attempt to define properties of interpretable models [61].

Doran et al. [33] proposed four notions of explainable AI—opaque systems, interpretable systems, comprehensible systems and truly explainable systems. They argue that people cannot hold accountable Artificial Intelligence systems without understanding of the decision making rationale of the system.

According to Miller, Howe and Sonenberg, [71], the problem with Explainable AI is that AI researchers do not use insights from the social sciences, especially philosophy, cognitive psychology/science, and social psychology. If they did, system would be more user-friendly and explanations would make more sense to an average user.

The author of this book could not agree more with this statement. Designing and developing Explainable Models requires putting humans in the center of the process. Only then it is possible to discover what level of explainability, interpretability, and transparency is needed in a specific use case and what all those qualities really mean from the user's perspective.

1.3 Recommender Systems

In the world of multimedia and the growing amount of electronic services, the choice of suitable products has become a troublesome and time-consuming issue. Due to the high availability of products or services, choosing a movie to watch, a mobile phone, or a car that meets all of the requirements consumes a considerable amount of time. As a result, time becomes the most significant value for a human being; moreover, its saving gains an appropriate value. The solution to this problem became recommendation systems (also called recommender systems or recommenders).

An appropriately tailored recommendation system for a specific industry allows for presenting an offer corresponding to the user preferences. In recent years, many recommendation systems for multimedia such as music, photos, or films have been created [5, 17, 39].

Recommendation systems have touched on other spheres of life like corporate media or the scientific world [24, 47, 48]. Such systems are becoming ubiquitous in every area of life. The preferences and values resulting from the users' behavior in the system allow for classifying them to a corresponding user model.

Despite information about the users' preferences and their context in a given recommendation system, one of the most valuable information is the rating of the presented product. The rating gives information to the system about how much the product is appropriate for a given user and should be treated as having an explicit impact [51].

A recommendation system is any system that offers items in a personalized way to a specific user or guides him to the product best suited to his profile [2, 26, 68]. In the literature, several techniques for building recommendations systems have been developed, including:

- Collaborative filtering—it is the most commonly implemented technique [36, 109]. Such systems recommend items by identifying other users with similar tastes. Recommendation for new items is based on user-user, user-item, or item-item similarities. The major problem with this technique is known under the name "cold start"—the system cannot draw any inferences for users or items about which it has not collected enough information.
- Content-based techniques—the recommender attempts to recommend items similar to those a given user preferred in the past [10, 89]. In this case, the recommen-

dations are based on information on the content of a given item, not on other users' opinions as in the case of collaborative filtering. These techniques are vulnerable to overfitting, but their great advantage is not needed for data on other users [125].
- Hybrid approach—it relies on a combination of many different recommendation methods [21]. The final goal of this approach is to obtain the most accurate list of predictions and, as a result, a more precise specification of the user's profile.

Without any doubt, almost all recommendation systems developed in the literature are focused on providing the highest accuracy of prediction (see e.g., [40, 41]). Consequently, recommender systems sacrifice their interpretability in favor of a correct rating. However, in many situations, we would like to understand why a certain recommendation has been made. Users have to be provided with explanations describing why particular items or services are recommended.

Rule-based AI recommender systems are knowledge-driven, while ML-based systems are data-driven; see e.g. [67].

Fuzzy sets and fuzzy rules have been applied in recommender systems but usually for the collaborative approach, e.g. [92, 96]. Recommenders based on collaborative filtering (see Sect. 1.3) suggest people to making choices taking into account the opinions and decisions of other people. A survey concerning the fuzzy approach in recommender systems can be found in [125]. The content-based recommenders are not often proposed with an application of Soft Computing methods (e.g. fuzzy logic, neural networks, genetic algorithms). However, there are some papers presented a fuzzy approach and related to the content-based recommendations, for example [122, 126].

The collaborative filtering has also been applied in recommenders based on neuro-fuzzy systems [81]. In this case, the ANFIS (Adaptive Network-based Fuzzy Inference System), introduced in [53], is employed.

Two most basic recommendation problems concern rating and item prediction. With regard to the rating prediction, there are users that rate items (e.g. movies, books, articles, or various devices). The rating is realized explicitly on some scale, e.g. using numbers 1 to 5, or stars, with meaning from least to most preferred item (product). Having such ratings, a recommender system should predict ratings of users for items that have not been rated yet. Speaking more formally, the following sets are considered: a set of users, U, a set of items, I, a set of ratings, e.g. $\{1, 2, 3, 4, 5\}$, and a set of triples, *(user, item, rating)*. In addition, a rating (loss) function that evaluates the difference between the predicted and actual rating; usually, the absolute error or square error is applied. For more examples and details about rating prediction, see e.g. [30, 63, 107, 112].

In the second case, when the item prediction is considered, there are no ratings—but (user, item) co-occurrences, e.g. users may view or buy some of the items. Thus, instead of the loss function, a score function is defined and employed in order to evaluate recommendations generated based on the co-occurrences of users and items (see e.g. [35]). According to [36], these two problems refer to explicit and implicit ratings, respectively.

A survey on applications of various recommendation systems can be found e.g. in [64]. Apart from the recommenders for multimedia [5, 17, 39], and others mentioned earlier, several examples of selected applications of different recommenders are listed as follows:

- Tourism: [16, 25, 31, 111]
- Education: [65, 105, 117, 118, 124]
- Web Pages: [2, 4, 11, 14, 22, 80, 104, 110]
- Social Systems: [40, 66, 116, 130]
- E-commerce: [76, 108, 121, 131]
- Health: [9, 44, 114, 120]
- Financial investments: [69, 78, 132]

Most recommender systems widely used in the domains listed above use a collaborative filtering approach, which works surprisingly well at a large scale with millions of users and ratings in the system. It can be observed in commercial platforms like Amazon.com, Booking.com, and Netflix.com.

Different methods are also applied in order to create recommendation systems; see e.g. [8, 54, 56, 85]. A survey on using ML algorithms is presented in [3, 91]. Soft computing methods in recommenders are considered and compared in [29]. As mentioned earlier, most of the literature concerns the collaborative approach [32, 36, 107, 115]. There are also papers about the application of deep learning with regard to recommendation systems [12, 129].

"Explainable Recommendation refers to the personalized recommendation algorithms that address the problem of why—they not only provide users with the recommendations, but also provide explanations to make the user or system designer aware of why such items are recommended." [127].

1.4 Interpretability of Machine Learning Models

In the last decade, we have witnessed tremendous progress in the performance of artificial intelligence (AI) systems. In the areas like game playing or image classification, AI techniques can achieve a very spectacular performance, sometimes even exceeding human-level skills. It is well known that the vast majority of these AI models works like the black box-models. However, in many applications, e.g. medical diagnosis or venture capital investment recommendations, it is essential to explain the rationale behind AI systems decisions or recommendations as discussed in Sect. 1.2.

This area of research has recently become a hot topic which attracted much attention among scientific community [6]. Among many scientific activities, for example, in 2019 special sessions devoted to this field have been organized during the following scientific conferences:

- FUZZ-IEEE 2019, New Orleans, USA, Advances on eXplainable Artificial Intelligence.
- IJCNN 2019, Budapest, Hungary, Explainable Machine Learning.

The 1st International Workshop on Explainable Recommendation and Search, EARS 2018, was co-located with the 41st International ACM SIGIR Conference on Research and Development in Information Retrieval (SIGIR 2018) held in Ann Arbor, Michigan, USA on July 12, 2018; see the Report [128].

It is also worth mentioning the 24th International Conference on Intelligent User Interfaces (ACM IUI'19), held in Los Angeles, USA, in March 2019, with a session on Recommender Systems, and the keynote lecture—DARPA's explainable artificial intelligence (XAI) program—by David Gunning.

The importance of XAI concerns many AI explainability problems e.g. in computer vision, deep learning, natural language processing, intelligent agents [1, 71, 82], and DARPA projects [42]. An important branch of AI research is the Explainable Machine Learning that is developed with regard to the explainable recommendation systems [127]. Of course, Machine Learning (ML) is widely applied in various practical recommendations; see e.g. [35]. Modern recommender systems are designed by the use of ML algorithms; for an excellent survey, a reader is referred to [91].

Many ML algorithms are treated as "black box" models. To develop a truly explainable AI system, it is a necessity to use interpretable models. In the literature, we can distinguish two approaches to achieve interpretability [43, 74]:

- Model-agnostic—techniques that can be applied to different types of ML algorithms.
- Model-specific—techniques that are applicable only for a single type or class of algorithm.

For instance, the LIME technique [97] is model agnostic and can be used to interpret nearly any set of ML inputs and ML predictions. The LIME algorithm can explain predictions of any classifiers or regressors by approximating it locally with an interpretable model. Although model-agnostic interpretability techniques are convenient, and in some ways, ideal, they often rely on surrogate models [70] or other approximations that can degrade the accuracy of the explanations they provide.

Model-specific interpretation techniques tend to use the model to be interpreter directly, leading to potentially more accurate explanations. Therefore, in this book, the model-specific approach is much more adequate to solving the challenge of developing truly explainable algorithms.

An essential criterion for explanations is interpretability. To be interpretable means provide qualitative understanding between input and output variables [97]. Explainable Artificial Intelligence (XAI) offers an explanation. It can come in many forms, e.g. text, visual, graph [127]. Interpretability is a quickly growing field in ML that is used in XAI. Interpretation methods apply ML models to produce relevant knowledge about relationships contained in data. Depending on the domain and context, this knowledge can be obtained in formats such as visualizations, natural language, or mathematical equations. A medical doctor, for example, in order to diagnose a patient, needs different information than an engineer [75]. Explainability and comprehensibility of AI are important requirements for intelligent systems deployed in real-world domains [123].

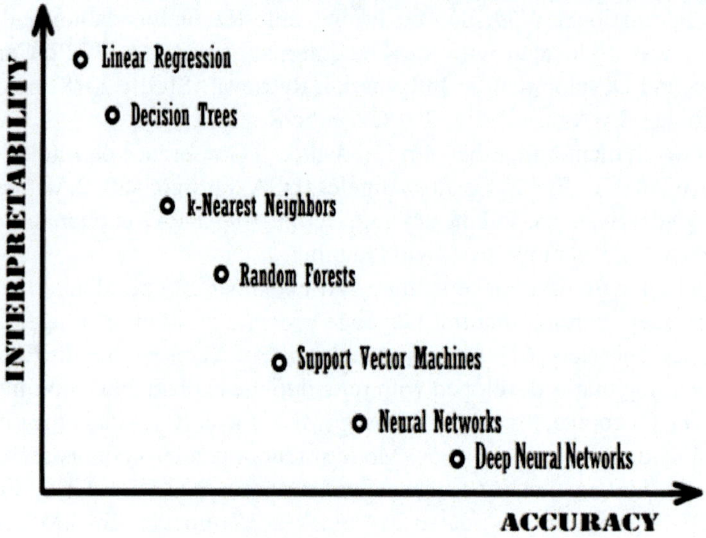

Fig. 1.5 Interpretability versus accuracy of ML models

There are several AI methods and techniques that could be considered for developing XAI recommender systems. Among them, rule-based systems are a natural choice. Typically, rule-based systems are, by definition, interpretable. Human beings can understand rules, therefore, it can be assumed that it is possible to develop explainable solutions using rules as a basis of interpretability.

However, there is a challenge of using data to generate rules automatically without losing too much accuracy. There is a known tradeoff between interpretability and accuracy; see e.g. [106].

Most of the researchers claim that accuracy is more important than interpretability. Especially in the case of choosing a movie to watch, it is easy to switch to another movie. It is not necessary for users to understand how the algorithm works as it is cheap to experiment with recommendations, and most users do not even care about explanations. In this case, interpretability might be a by-product if the algorithm provides it anyway or even a superfluous luxury. However, depending on the domain and particular objectives, interpretability might be a necessity. It is unlikely to happen that one would make a critical decision based on the recommendation provided by an algorithm without an explanation and argumentation. Taking into consideration more critical problems, e.g. matchmaking or venture capital investments, it turns out that some kind of understanding of how the model works is crucial to its users.

Figure 1.5, portrayed based on [98], presents few AI methods (applied as ML models) from the perspective of interpretability vs. accuracy tradeoff.

Let us notice that the Linear Regression (see e.g. [15, 79]) is interpretable because the model is represented by the regression function while Neural Networks (see e.g. [86, 133]) are treated as a "black box" model. Decision Trees can be interpreted by corresponding logical IF-THEN rules [23, 93] but SVM (Support Vector Machines) is a method related to neural networks [27, 49, 58], however more interpretable. Although Random Forests [19, 60] are composed of decision trees, the interpretability is much more difficult since the model is more complex. The k-Nearest Neighbors [7] is a very well known and popular method in ML, and intuitively can be interpretable with regard to the neighboring data viewed as similar.

It is worth mentioning that some attempts have been made in order to extract logical IF-THEN rules from hidden knowledge of the "black box" models realized by neural networks; see e.g. [34, 46]. This is important from the XAI point of view. It should be added that the backpropagation algorithm of neural network learning (see e.g. [133]) can give a little interpretability. However, Deep Neural Networks [13, 50, 57], with a large number of layers and weights, is much more complex and hard to interpret.

The rule-based methods are not included in Fig. 1.5, since—as mentioned earlier—such models are interpretable (see also [62]). The knowledge of the form of IF-THEN rules have been incorporated in expert systems [20, 52, 95], almost from the beginning of Artificial Intelligence [100]. Expert systems, successfully applied in many areas such as medicine, industry, geology, computer science, and more, are equipped with the so-called "explanation facilities" in order to explain the inference and systems' decisions. Moreover, as a matter of fact, the rule-based systems are not ML methods since, in this case, the system knowledge is known from experts, and does not need to be acquired by learning.

Other rule-based methods, well known in AI, with many practical applications, are fuzzy systems (see e.g. [55, 90]) that use knowledge in the form of fuzzy IF-THEN rules, and intelligent systems with the knowledge formulated as logical IF-THEN rules that include rough sets [87, 88]. Both kinds of systems model uncertainty in their knowledge bases. The main assumption is that the rules are provided by experts. However, it is possible to generate the IF-THEN rules from data. This means that the systems can acquire the rules by learning. As a matter of fact, such systems are constructed as a combination of fuzzy or rough set systems with neural networks. Thus, we consider neuro-fuzzy systems (see e.g. [77, 101, 102]), and also rough-neuro-fuzzy systems (see e.g. [28, 83]).

1.5 The Content and Main Results of the Book

This book contains five chapters. In Chap. 2, the background of neuro-fuzzy systems is briefly outlined, along with issues of interpretability and explainability. Moreover, two commonly used methods for fuzzy rules generation are presented. Namely, techniques of Wang-Mendel [119] and Nozaki-Ischibuchi-Tanaka [84], applied as a starting point for designing new explainable recommender systems, described in Chaps. 3 and 4. In Chap. 3, several new designs of recommender systems are presented. They are marked as "Recommender A", "Recommender B", and "Recommender C". The recommenders differ in nominal values encoding and in optimization methods. It should be emphasized that converting nominal values of features (attributes), e.g., "movie genres" or "actors" (for the movie data), is not a trivial problem, and in the world literature, there is a lack of effective solutions. In Chap. 3, two new methods are presented and formally described: (a) in Sect. 3.1.1—a method used for the "Recommender A" and for "Recommender B", (b) in Sect. 3.3.1—a method used for "Recommender C".

The "Recommender A", proposed in Sect. 3.1, presents an application of the "Zero-Order Takagi-Sugeno-Kang" (ZO-TSK) method to explainable recommender systems. This method is based on the Wang-Mendel (WM) and Nozaki-Ishibuchi-Tanaka (NIT) techniques for fuzzy rule generation, and it is best suited to predict users' ratings. The model is optimized by the use of the "Grey Wolf Optimizer" [37, 72], without affecting the interpretability. The performance of the methods is illustrated on the MovieLens 10M dataset [45].

The proposed approach allows achieving high accuracy with a reasonable number of interpretable fuzzy rules. The use of the ZO-TSK and optimization of consequent singletons allows obtaining significant improvements to the results. Experiments prove that applying the "Grey Wolf Optimizer" to train the model gives better accuracy without losing the interpretability of the system. The ZO-TSK fuzzy system can be effectively used as a content-based recommender system that provides accurate results with interpretability, transparency, and explainability.

The "Recommender B", described in Sect. 3.2, is proposed using a novel approach to design explainable recommender systems. It is based on the WM algorithm for fuzzy rule generation. In Sect. 3.2.1, a method for learning and reduction of the fuzzy recommender is presented. Three criteria, including the Akaike information criterion [18], are used for evaluating an optimal balance between the recommender accuracy and interpretability. Simulation results verify the effectiveness of the presented recommender system and illustrate its performance on the MovieLens 10 M dataset.

The explainability of the proposed recommender is assured due to:

- Interpretable fuzzy rules with fuzzy sets as linguistic values of attributes describing items (objects) and users' preferences—fuzzy sets with semantic meanings.
- Incorporation of rule weights into the neuro-fuzzy system; the weights interpreted with regard to rule importance.
- The reduction of fuzzy rules that makes the rule base simpler, and therefore easier to produce explainable recommendations.

It is worth emphasizing that this approach leads to a moderate number of interpretable fuzzy rules, and in consequence, significantly facilities the explanation of the recommender system. Moreover, the use of the Akaike information criterion, as well as the final prediction error and the Schwartz criterion (see Sect. 3.3.2), allow solving the problem of the compromise between the system error and the number of rules.

The "Recommender C", described in Sect. 3.3, is based on the novel method for nominal attribute encoding. Several flexibility parameters—subject to learning—are incorporated into their construction, allowing systems to represent patterns encoded in data better. The learning process does not affect the initial interpretable form of fuzzy recommenders' rules. Using the Akaike Information criterion allows evaluating the trade-off between the number of rules and interpretability, which is crucial to provide proper explanations for users.

A novelty and characteristics of the "Recommender C" are summarized as follows:

- Explainability of recommender systems is assured by generating a moderate number of interpretable fuzzy IF-THEN rules based on input-output data.
- A new method, well justified by mathematical statistics, for transforming nominal values into a numerical form is presented. This is the first mathematically justified technique for such transformation and allows representing nominal values, e.g., movie genres, keywords or actors, in a fuzzy system designed based on the input-output observations.
- Membership functions of fuzzy sets in the IF-THEN rules are fixed during the learning process to preserve their linguistic interpretation.
- Two groups of flexibility parameters are incorporated into the construction of the architecture of proposed fuzzy recommender systems: Rules' weights describing the importance of the rules, and parameters describing a parameterized triangular T-norm (see e.g. [103]) used in fuzzy systems to connect antecedents and consequents of the particular rules. Learning of these parameters significantly improves the recommender system performance allowing to perfectly represent the patterns encoded in data without losing rules' interpretability.
- A simple yet effective reduction mechanism is proposed, allowing to reduce the number of initial fuzzy rules.
- The Akaike information criterion is applied to solve the problem of the compromise between the recommender systems' error and the number of rules.

Chapter 4 describes the assumptions of the recommendation system, in which historical examples are available from only one class (one-class classification). It was based on real data and a problem was defined in which the explanation is critical. While in recommending movies, the consequences of a bad recommendation are low, in stock market investments, the consequences of making a wrong decision can be very costly. For this reason, in the recommender system for investment advisors, recommendations alone are not enough. It is necessary to explain why a given recommendation is the best and how the model came up with it. To create an explainable recommendation system for investment advisors, the author used publicly available data which were enriched with data provided by companies listed on the stock markets. As a result, 21 values were obtained for each transaction that corresponded to the criteria for making investment decisions. Based on this data, the membership functions of fuzzy sets were statistically determined, which were used to generate the explanation understandable to the end user. However, to ensure the accuracy and adaptability of the system, for each historical transaction, an algorithm created a fuzzy set in a multidimensional space. Instead of creating a model that generalizes the decision-making process by trying to fit as much of the sample data as possible, in the innovative approach presented in chapter four, all available data form a model, and pattern recognition is based on the similarity of historical examples to the new object being considered for recommendation (in this case, a stock described by 21 features and available on the stock markets at any given time). What distinguishes the method from the other available approaches is the use of the degree of activation of the rule (and how it is determined) as a measure of the neighborhood, and thus the similarity of the objects.

The summary and final remarks, in Chap. 5, conclude this book.

Chapters 3 and 4 contain the novel contribution of the author. Referring to Sect. 1.3, recommenders described in Chap. 3 should be viewed with regard to the rating prediction (explicit ratings) while Chap. 4 presents the recommendation system that works based on the implicit ratings (item prediction). In addition, the former systems solve regression or multi-class classification problems while the latter recommender is a one-class classifier.

References

1. Adadi, A., Berrada, M.: Peeking inside the black-box: a survey on Explainable Artificial Intelligence (XAI). IEEE Access **6**, 52138–52160 (2018)
2. Adeniyi, D.A., Wei, Z., Yongquan, Y.: Automated web usage data mining and recommendation system using K-Nearest Neighbor (KNN) classification method. Appl. Comput. Inf. **12**(1), 90–108 (2016)
3. Aggarwal, C.C.: Recommender Systems. Springer, Berlin (2016)
4. Aggarwal, S.: Web Page Recommender System Based on Data Mining: Hidden Markov Model to Explore the Unknown, ed. by V. Mangan. Lap Lambert Academic Publishing (2016)
5. Aizenberg, N., Koren, Y. Somekh, O.: Build your own music recommender by modeling internet radio streams. In: Proceedings of the 21st International Conference on World Wide Web (WWW '12), pp. 1–10. ACM Press. New York (2012)
6. Alonso, J.M., Castiello, C., Mencar, C.: A bibliometric analysis of the explainable artificial intelligence research field. In: Medina, J. et al. (ed.) Information Processing and Management of Uncertainty in Knowledge-Based Systems. Theory and Foundations, pp. 3–5 (2018)
7. Altman, N.S.: An introduction to kernel and nearest neighbor nonparametric regression. Am. Stat. **46**(3), 175–185 (1992)
8. Anaissi, A., Goyal, M.: SVM-based association rules for knowledge discovery and classification. In: The 2nd Asia-Pacific World Congress on Computer Science and Engineering (APWC on CSE), pp. 1–5 (2015)
9. Archenaa, J., Anita, M.: Health recommender system using big data analytics. J. Manag. Sci. Bus. Intell. **2**(2), 17–23 (2017)
10. Bagher, R.C., Hassanpour, H., Mashayekhi, H.: User trends modeling for a content-based recommender system. Exp. Syst. Appl. **87**, 209–219 (2017)
11. Barot, M., Wandra, K.H., Patel, S.B.: Web usage data based web page recommender system. Int. J. Eng. Dev. Res. (IJEDR) **5**(2), 1769–1775 (2017)
12. Batmaz, Z., Yurekli, A., Bilge, A., Kaleli, C.: A review on deep learning for recommender systems: challenges and remedies. Artif. Intell. Rev. **52**(1), 1–37 (2019)
13. Bengio, Y.: Learning deep architectures for AI. Found. Trends Mach. Learn. **2**(1), 1–127 (2009)
14. Bhoomika, A.P., Selvarani, R.: A Survey on web page recommender systems. Presented at the (2019)
15. Bishop, C.M.: Pattern Recognition and Machine Learning. Springer, Berlin (2006)
16. Borràs, J., Moreno, A., Valls, A.: Intelligent tourism recommender systems: a survey. Exp. Syst. Appl. **4**(16), 7370–7389 (2014)

17. Bourke, S., McCarthy, K., Smyth, B.: The social camera: a case-study in contextual image recommendation. In: Proceedings of the 16th International Conference on Intelligent User Interfaces, pp. 13–22 (2011)
18. Bozdogan, H.: Model selection and Akaike's information criterion (AIC): the general theory and its analytical extensions. Psychometrika **52**(3), 345–370 (1987)
19. Breiman, L.: Random forests. Mach. Learn. **45**, 5–32 (2001)
20. Buchanan, B.G., Shortliffe, E.H.: Rule-Based Expert Systems: The MYCIN Experiments of the Stanford Heuristic Programming Project. Addison-Wesley (1985)
21. Burke, R.: Hybrid recommender systems: survey and experiments. User Model. User-Adapt. Interaction **12**(4), 331–370 (2002)
22. Burke, R.: Hybrid web recommender systems. In: Brusilovsky, P., Kobsa, A., Nejdl, W. (eds.) The Adaptive Web: Methods and Strategies of Web Personalization, pp. 377–408. Springer, Berlin Heidelberg (2007)
23. Chiang, D.-A., Chen, W., Wang, Y.-F., Hwang, L.-J.: Rules generation from the decision tree. J. Inf. Sci. Eng. **17**(2), 325–339 (2001)
24. Chin, A., Xu, B., Wang, H.: Who should I add as a 'friend'?: a study of friend recommendations using proximity and homophily. In: Proceedings of the 4th International Workshop on Modeling Social Media (MSM'13), pp. 7:1–7:7 (2013)
25. Codina, V.: A Recommender System for the Semantic Web: Application in the Tourism Domain. Scholar Press (2012)
26. Conforti, R., de Leoni, M., La Rosa, M., van der Aalst, W.M.P., ter Hofstede, A.H.M.: A recommendation system for predicting risks across multiple business process instances. Decision Support Syst. **69**, 1–19 (2015)
27. Cortes, C., Vapnik, V.: Support-vector networks. Mach. Learn. **20**(3), 273–297 (1995)
28. Cpałka, K., Nowicki, R., Rutkowski, L.: Rough-neuro-fuzzy systems for classification. In: IEEE Symposium on Foundations of Computational Intelligence (FOCI), pp. 1–8 (2007)
29. Das, S., Mishra, B.S.P., Mishra, M.J., Mishra, S., Moharana, S.C.: Soft-computing based recommendation system: a comparative study. Int. J. Innov. Technol. Exploring Eng. (IJITEE) **8**(8), 131–139 (2019)
30. Deshpande, M., Karypis, G.: Item-based top-N recommendation algorithms. ACM Trans. Inf. Syst. **22**(1), 143–177 (2004)
31. Dietz, L.W., Myftija, S., Wörndl, W.: Designing a conversational travel recommender system based on data-driven destination characterization. RecTour **2019**, 17 (2019)
32. Divya, N., Sandhiya, S., Liz, A.S., Gnanaoli, P.: A collaborative filtering based recommender system using rating prediction. Int. J. Pure Appl. Math. **119**(10), 1–7 (2018)
33. Doran, D., Schulz, S., Besold, T.R.: What Does Explainable AI Really Mean? A New Conceptualization of Perspectives (2017)
34. Duch, W., Setiono, R., Żurada, J.M.: Computational intelligence methods for rule-based data understanding. Proc. IEEE **92**(5), 771–805 (2004)
35. Duning, T., Friedman, E.: Practical Machine Learning: Innovations in Recommendation. O'Reilly Media, Inc. (2014)
36. Ekstrand, M.D., Riedl, J.T., Konstan, J.A.: Collaborative filtering recommender systems. Found. Trends® Human-Comput. Interaction **4**(2), 81–173 (2011)
37. Faris, H., Aljarah, I., Al-Betar, M.A., Mirjalili, S.: Grey wolf optimizer: a review of recent variants and applications. Neural Comput. Appl. **3**(2), 413–435 (2018)
38. Fjeld, J., Achten, N., Hilligoss, H., et al.: Principled Artificial Intelligence: Mapping Consensus in Ethical and Rights-Based Approaches to Principles for AI (2020)
39. Gantner, Z., Rendle, S., Schmidt-Thieme, L.: Factorization models for context-/time-aware movie recommendations. In: Proceedings of the Workshop on Context-Aware Movie Recommendation (CAMRa '10), pp. 14–19 (2010)
40. Gedikli, F.: Recommender Systems and the Social Web: Leveranging Tagging Data for Recommender Systems. Springer Vieweg, Dortmund, Germany (2012)
41. Gunawardana, A., Shani, G.: A survey of accuracy evaluation metrics of recommendation tasks. J. Mach. Learn. Res. **10**, 2935–2962 (2009)

42. Gunning, D., Aha, D.: DARPA's Explainable Artificial Intelligence (XAI) program. AI Mag. **40**(2), 44–58 (2019)
43. Hall, P., Gill, N.: An Introduction to Machine Learning Interpretability: An Applied Perspective on Fairness, Accountability, Transparency, and Explainable AI. O'Reilly Media, Inc. (2018)
44. Hao, F., Blair, R.: A comparative study: classification vs. user-based collaborative filtering for clinical prediction. BMC Med. Res. Methodol. **16**(1): 172 (2016)
45. Harper, F.M., Konstan, J.A.: The MovieLens datasets: history and context. ACM Trans. Interactive Intell. Syst. **5**(4), 19:1-19:19 (2015)
46. Hayashi, Y., Setiono, R., Azcarraga, A.: Neural network training and rule extraction with augmented discretized input. Neurocomputing **207**, 610–622 (2016)
47. He, J., Nie, J.-Y., Lu, Y., Zhao, W.X.: Position-aligned translation model for citation recommendation. In: Calderón-Benavides, L., González-Caro, C., Chávez, E., Ziviani N. (eds.) String Processing and Information Retrieval, pp. 251–263 (2012)
48. He, Q., Kifer, D., Pei, J., Mitra, P., Giles, C.L.: Citation recommendation without author supervision. In: Proceedings of the Fourth ACM International Conference on Web Search and Data Mining, pp. 755–764 (2011)
49. Hearst, M.A., Dumais, S.T., Osuna, E., Platt, J., Scholkopf, B.: Support vector machines. IEEE Intell. Syst. Appl. **13**(4), 18–28 (1998)
50. Hinton, G.E., Osindero, S., Teh, Y.-W.: A fast learning algorithm for deep belief nets. Neural Comput. **18**(7), 1527–1554 (2006)
51. Isinkaye, F.O., Folajimi, Y.O., Ojokoh, B.A.: Recommendation systems: principles, methods and evaluation. Egyptian Inform. J. **16**(3), 261–273 (2015)
52. Jackson, P.: Introduction to Expert Systems, 3rd edn. Addison-Wesley Longman Publishing Co. Inc., Boston, MA, USA (1998)
53. Jang, J.S.R.: ANFIS: adaptive-network-based fuzzy inference system. IEEE Trans. Syst. Man Cybern. **23**(3), 665–685 (1993)
54. Kim, K., Ahn, H.: A recommender system using GA K-means clustering in an online shopping market. Exp. Syst. Appl. **34**(2), 1200–1209 (2008)
55. Klir, G. J., Yuan, B. (eds.): Fuzzy Sets, Fuzzy Logic and Fuzzy Sytems: Selected Papers by Lotfi A. Zadeh. Adv. Fuzzy Syst. Appl. Theory **6** (1996)
56. Kunaver, M., Požrl, T.: Diversity in recommender systems—a survey. Knowl.-Based Syst **123**, 154–162 (2017)
57. LeCun, Y., Bengio, Y.: Convolutional networks for images, speech, and time-series. Handbook Brain Theory Neural Netw. **3361**(10) (1995)
58. Lee, M.-C., To, C.: Comparison of support vector machine and back propagation neural network in evaluating the enterprise financial distress. Int. J. Artif. Intell. Appl. (IJAIA) **1**(3) (2010)
59. Lewis, D.: On the Plurality of Worlds. Blackwell, Oxford (1986)
60. Liaw, A., Wiener, M.: Classification and regression by randomForest. R News **2**(3), 18–22 (2002)
61. Lipton, Z.C.: The mythos of model interpretability. ACM Queue **61**(10), 36–43 (2018)
62. Liu, H., Gegov, A., Cocea, M.: Rule based networks: an efficient and interpretable representation of computational models. J. Artif. Intell. Soft Comput. Res. **7**(2), 111–123 (2017)
63. Lops, P., de Gemmis, M., Semeraro, G.: Content-based recommender systems: state of the art and trends. In: F. Richi et al. (eds.) Recommender Systems Handbook, pp. 73–105. Springer, Berlin (2011)
64. Lu, J., Wu, D., Mao, M., Wang, W., Zhang, G.: Recommender system application: a survey. Decision Sup. Syst. **74**, 12–32 (2015)
65. Manouselis, N., Drachsler, H., Verbert, K., Santos, O.C.: Recommender Systems for Technology Enhanced Learning: Research Trends and Applications. Springer, Berlin (2014)
66. Maringho, L.B., Hotho, A., Jaschke, R., Nanopoulos, A., Rendle, S., Schmidt-Thieme, L., Stumme, G., Symeonidis, P.: Recommender Systems for Social Tagging Systems. Springer, Berlin (2012)
67. Mary, J.: Data-Driven Recommender Systems: Sequences of recommendations. Université de Lille, Artificial Intelligence (2015)

68. Meehan, K., Lunney, T., Curran, K., McCaughey, A.: Context-aware intelligent recommendation system for tourism. In: 2013 IEEE International Conference on Pervasive Computing and Communications Workshops (PERCOM Workshops), pp. 328–331 (2013)
69. Meier, A., Portmann, E., Teran, L. (eds.): Applying Fuzzy Logic for the Digital Economy and Society. Springer, Berlin (2019)
70. Messalas, A., Kanellopoulos, Y., Makris, C.: Model-agnostic interpretability with Shapley values. In: 2019 10th International Conference on Information, Intelligence, Systems and Applications (IISA), pp. 1–7 IEEE (2019)
71. Miller, T.: Explanation in Artificial Intelligence: insights from the social sciences. Artif. Intell. **267**, 1–38 (2017)
72. Mirjalili, S., Mirjalili, S.M., Lewis, A.: Grey Wolf optimizer. Adv. Eng. Softw. **69**, 46–61 (2014)
73. Mishra, S., Sahoo, S., Mishra, B.K.: Neuro-fuzzy models and applications. In: Emerging Trends and Applications in Cognitive Computing. IGI Global, pp. 78–98 (2019)
74. Molnar, C.: Interpretable Machine Learning: A Guide for Making Black Box Models Explainable. Lean Publishing (2018)
75. Murdoch, W.J., Singh, C., Kumbier, K., Abbasi-Asl, R., Yu, B.: Interpretable machine learning: definitions, methods, and applications. Proc. Natl. Acad. Sci. U. S. A. **116**(44), 2207–22080 (2019)
76. Narayan, V., Mehta, R.K., Rai, M., Gupta, A., Verma, S., Patel, A., Yadav, S.: E-commerce recommendation method based on collaborative filtering technology. Int. J. Current Eng. Technol. **7**(3) (2017)
77. Nauck, D., Klawonn, F., Kruse, R.: Foundations of Neuro-Fuzzy Systems. Wiley, New York (1997)
78. Nayak, B., Ojha, R.K., Subbarao, P.S., Bath, V.: Machine Learning finance: application of Machine Learning in collaborative filtering recommendation system for financial recommendations. Int. J. Recent Technol. Eng. (IJRTE) **8**(1), 905–909 (2019)
79. Neter, J., Wasserman, W., Kutner, M.H.: Applied Linear Regression Models, 4th edn. McGraw-Hill Education - Europe, London, United States (2003)
80. Nguyen, T.T.S., Lu, H.Y., Lu, J.: Web-page recommendation based on web usage and domain knowledge. IEEE Trans. Knowl. Data Eng. **26**(10), 2574–2587 (2014)
81. Nilashi, M. bin Ibrahim, O. Ithin, N., Sarmin, N.H.: A multi-criteria collaborative filtering recommender system for the tourism domain using Expectation Maximization (EM) and PCA-ANFIS. Electron. Commerce Res. Appl. **14**(6), 542–562 (2015)
82. Nott, G.: Explainable Artificial Intelligence: cracking open the black box of AI. Computer World. https://www.computerworld.com.au/article/617359/
83. Nowicki, R.: Rough neuro-fuzzy structures for classification with missing data. IEEE Trans. Syst. Man Cybern. Part B (Cybernetics) **39**(6), 1334–1347 (2009)
84. Nozaki, K., Ishibuchi, H., Tanaka, H.: A simple but powerful heuristic method for generating fuzzy rules from numerical data. Fuzzy Sets Syst. **86**(3), 251–270 (1997)
85. Park, D.H., Kim, H.K., Choi, I.Y., Kim, J.K.: A literature review and classification of recommender systems research. Exp. Syst. Appl. **39**(11), 10059–10072 (2012)
86. Patterson, D.W.: Artificial Neural Networks: Theory and Applications. Prentice Hall (1996)
87. Pawlak, Z.: Rough sets. Int. J. Comput. Inf. Sci. **11**(5), 341–356 (1982)
88. Pawlak, Z.: Rough Sets: Theoretical Aspects of Reasoning about Data. Kluwer Academic Publishers (1991)
89. Pazzani, M.J., Billsus, D.: Content-based recommendation systems. In: P. Brusilovsky, A. Kobsa, W. Nejdl (eds.) The Adaptive Web, vol. 4321, pp. 325–341. Springer, LNCS (2007)
90. Pedrycz, W.: Fuzzy Control and Fuzzy Systems. Wiley, New York, NY, USA (1993)
91. Portugal, I., Alencar, P., Cowan, D.: The use of machine learning algorithms in recommender systems: a systematic review. Exp. Syst. Appl. **97**, 205–227 (2018)
92. Prasad, M., Liu, Y.-T., Li, D.-L., Lin, C.-T., Shah, R.R., Kaiwartya, O.P.: A new mechanism for data visualization with TSK-type preprocessed collaborative fuzzy rule based system. J. Artif. Intell. Soft Comput. Res. **7**(1), 33–46 (2017)

93. Quinlan, J.R.: Generating production rules from decision trees. In: Proceedings of the 10th International Joint Conference on Artificial Intelligence (IJCAI), pp. 304–307 (1987)
94. Rathi, S.: Generating counterfactual and contrastive explanations using SHAP (2019). arXiv:1906.09293
95. Ravuri, M., Kannan, A., Tso, G.J., Amatriain, X.: Learning from the experts: from expert systems to machine-learned diagnosis models. Proc. Mach. Learn. Res. **8**, 1–16 (2018)
96. Reformat, M.Z., Yager, R.R.: Suggesting recommendations using Pythagorean fuzzy sets illustrated using Netflix movie data. In: Laurent, A., et al. (eds.) International Conference on Information Processing and Management of Uncertainty in Knowledge-Based Systems (IPMU 2014), Part I, CCIS 442, pp. 546–556. Springer International Publishing, Switzerland (2014)
97. Ribeiro, M.T., Singh, S., Guestrin, C.: Why should I trust you?: explaining the predictions of any classifier. In: Proceedings of the 22nd ACM SIG KDD International Conference on Knowledge Discovery and Data Mining, pp. 1135–1144 (2016)
98. Rodrigues, J.: This New Google Technique Help Us Understand How Neural Networks are Thinking. Towards Data Science. www.towardsdatascience.com
99. Rudin, C.: Stop explaining black box machine learning models for high stakes decisions and use interpretable models instead. Nat. Mach. Intell. **1**(5), 206–2015 (2019)
100. Russel, S., Norvig, P.: Artificial Intelligence: A Modern Approach. 3rd edn. Prentice Hall. Series in Artificial Intelligence (2010)
101. Rutkowska, D.: Neuro-Fuzzy Architectures and Hybrid Learning. Physica-Verlag. Springer, Heidelberg, New York (2002)
102. Rutkowski, L.: Flexible Neuro-Fuzzy Systems: Structures. Kluwer Academic Publisher, Learning and Performance Evaluation (2004)
103. Rutkowski, L.: Computational Intelligence: Methods and Techniques. Springer, Berlin (2008)
104. Suguna, R., Sharmila, D.: An efficient web recommendation system using collaborative filtering and pattern discovery algorithms. Int. J. Comput. Appl. **70**(3), 37–44 (2013)
105. Santos, O.C., Boticario, J.G.: Educational Recommender Systems and Technologies: Practices and Challenges. IGI Global (2012)
106. Sarkar, S., Weyde, T., Garcez, A., Slabaugh, G.G., Dragicevic, S., Percy, C.: Accuracy and interpretability trade-offs in machine learning applied to safer gambling. CEUR Workshop Proc. **1773**, (2016)
107. Sarwar, B., Karypis, G., Konstan, J., Riedl, J.: Item-based collaborative filtering recommendation algorithms. In: WWW '01 Proceedings of the 10th International Conference on World Wide Web, pp. 285–295 (2001)
108. Schafer, J.B., Konstan, J.A., Riedl, J.: E-commerce rcommendation applications. Data Mining Knowl. Discovery **5**(1–2), 115–153 (2001)
109. Schafer, J.B., Frankowski, D., Herlocker, J., Sen, S.: Collaborative filtering recommender systems". In: Brusilovsky, P., Kobsa, A., Nejdl, W. (eds.) The Adaptive Web: Methods and Strategies of Web Personalization, pp. 291–324. Springer, Berlin, Heidelberg (2007)
110. Shani, G., Chickering, M., Meek, C.: "Mining recommendations from the Web". In: Proceedings of the 2008 ACM Conference on Recommender Systems (RecSys'08), pp. 35–42 (2008)
111. Sharda, N. (ed.): Tourism Informatics: Visual Travel Recommender Systems, Social Communities, and User Interface Design. IGI Global, USA (2010)
112. Sidana, S.: Recommendation systems for online advertising. Ph.D. thesis. Computers and Society Université Grenoble Alpes (2018)
113. Sousa, P.V. de C.: Fuzzy neural networks and neuro-fuzzy networks: a review the main techniques and applications used in the literature. Appl. Soft Comput. J. **92**, 106275 (2020)
114. Stark, B., Knahl, C., Aydin, M., Elish, K.: A literature review on medicine recommender systems. Int. J. Adv. Comput. Sci. Appl. (IJACSA) **10**(8) (2019)
115. Su, X., Khoshgoftaar, T.M.: A survey of collaborative filtering techniques. Adv. Artif. Intell. **2009**(4) (2009)
116. Symeonidis, P., Ntempos, D., Manolopoulos, Y.: Recommender Systems for Location-Based Social Networks. Springer, Berlin (2014)

117. Toledo, R.Y., Mota, Y.C., Martinez, L.: A recommender system for programming online judges using fuzzy information modeling. Informatics **5** (2018)
118. Tsuji, K., Yoshikane, F., Sato, S., Itsumura, H.: Book recommendation using machine learning methods based on library loan records and bibliographic information. In: 2014 IIAI 3rd International Conference on Advanced Applied Informatics, pp. 76–79 (2014)
119. Wang, L.-X., Mendel, J.M.: Generating fuzzy rules by learning from examples. IEEE Trans. Syst. Man Cybern. **22**(6), 1414–1427 (1992)
120. Vishwajith, V., Kaviraj, S., Vasanth, R.: Hybrid recommender system for therapy recommendation. Int. J. Adv. Res. Comput. Commun. Eng. (IJARCCE) **8**(1), 78–84 (2019)
121. Ya, L.: The comparison of personalization recommendation for e-commerce. Phys. Procedia **25**, 475–478 (2012)
122. Yager, R.: Fuzzy logic methods in recommender systems. Fuzzy Sets Syst. **136**(2), 133–149 (2003)
123. Yampolskiy, R.V.: Unexplainability and incomprehensibility of Artificial Intelligence (2019). arXiv:1907.03869
124. Yang, W.-S., Lin, Y.-R.: A task-focused literature in recommender systems for digital libraries. Online Inf. Rev. **37**(4), 581–601 (2013)
125. Yera, R., Martinez, L.: Fuzzy tools in recommender systems: a survey. Int. J. Comput. Intell. Syst. **10**(1), 776–803 (2017)
126. Zenebe, A., Norcio, A.: Representation, similarity measures and aggregation methods using fuzzy sets for content-based recommender systems. Fuzzy Sets Syst. **160**(1), 76–94 (2009)
127. Zhang, Y., Chen, X.: Explainable recommendation: a survey and new perspectives, 11192 (2018). arXiv: 1804
128. Zhang, Y., Zhang, Y., Zhang, M.: Report on EARS18: 1st international workshop on explainable recommendation and search. ACM SIGIR Forum **52**(2), 125–131 (2018)
129. Zhang, S., Yao, L., Sun, A., Tay, Y.: Deep learning based recommender system: a survey and new perspectives. ACM Comput. Surv. **1**(1) (2018)
130. Zhao, K., Pan, L.: A Machine Learning based trust evaluation framework for online social networks. In: 2014 IEEE 13th International Conference on Trust, Security and Privacy in Computing and Communications, pp. 69–74 (2014)
131. Zhao, Q., Zhang, Y., Friedman, D., Tan, F.: E-commerce recommendation with personalized promotion. In: Proceedings of the 9th ACM Conference on Recommender Systems, pp. 219–226 (2015)
132. Zibriczky, D.: Recommender systems meet finance: a literature review. In: Proceedings of the 2nd International Workshop on Personalization and Recommendation Systems in Financial Services (FINREC 2016), pp. 3–10 (2016)
133. Żurada, J.M.: Introduction to Artificial Neural Systems. West Publishing Company (1992)

Chapter 2
Neuro-Fuzzy Approach and Its Application in Recommender Systems

2.1 Neuro-Fuzzy Systems as Recommenders

As mentioned in Sect. 1.4, a combination of fuzzy systems [22, 24, 37, 58] with neural networks [3, 12, 36, 59] allows introducing the learning ability of neural networks to fuzzy systems that inference based on fuzzy IF-THEN rules; for details, see also e.g. [18, 19, 23, 27, 40, 57]. Fuzzy systems work as expert systems [6, 15, 30, 43], and are interpretable. The explanation is possible based on the rules. Neural networks are "black box" models that acquire knowledge from data through the use of a learning algorithm.

Neuro-fuzzy systems [8, 32, 40, 41, 44] are proposed as recommenders in this book. However, the neuro-fuzzy systems cannot be directly applied as recommenders. It should be emphasized that neural networks and neuro-fuzzy systems process numerical data. This means that the input and output values of such systems are numbers. Recommendation systems can produce numbers as output values but also other types, e.g. logical or categorical values. Their input values can also be numerical but often are nominal (not numerical) values of various attributes of objects to be recommended. Therefore, the main issue is to use an algorithm of encoding the nominal values into numerical ones.

In Chap. 3, new neuro-fuzzy recommenders are presented (denoted as A, B, C), and for each of them a feature encoding procedure is proposed. However, the novel recommendation system described in Chap. 4 uses numerical values of attributes, so it does not need feature encoding algorithms. It should be emphasized that this recommender differs from those applied in Chap. 3, and incorporates other new ideas. Of course, a type of the recommender depends on the recommendation problems and datasets.

T. Rutkowski, *Explainable Artificial Intelligence Based on Neuro-Fuzzy Modeling with Applications in Finance*, Studies in Computational Intelligence 964,
https://doi.org/10.1007/978-3-030-75521-8_2

2.2 Fuzzy IF-THEN Rules and Learning Ability of the Recommenders

It is worth noticing that recommendations are usually proposed as a result of data analysis. Based on datasets that contain information about previous recommendations of various items to different users, along with users' opinions about the recommended products, it is possible to propose new appropriate recommendations. Therefore, a dataset of this kind is employed in order to construct a recommender, using a data mining algorithm, which is a Machine Learning (ML) method of AI; see Chap. 1, and e.g. [4, 10, 55].

Referring to Sect. 1.4, there are two main approaches, the rule-based systems, and machine learning models—as portrayed in Fig. 1.5—with different ratios of interpretability versus accuracy. Of course, the learning-based methods are less interpretable. However, it is difficult to formulate expert rules without domain knowledge. If we had the rules, we would apply a fuzzy system as a recommender. Otherwise, we can acquire the knowledge—represented as a rule base—from data. This means that the learning ability is introduced to the recommender. The learning algorithm can be similar to that employed in neural networks, such as the backpropagation (see e.g. [59]). As a matter of fact, we assume a general form of the IF-THEN rules, and then learn parameters of the rules, analogously to the neural networks' weights learning; see e.g. [26, 40, 52].

The fuzzy rules, denoted as R^k, for $k = 1, 2, \ldots, N$, are usually of the following form:

$$\textbf{IF } x_1 \text{ is } A_1^k \text{ AND } x_2 \text{ is } A_2^k \text{ AND} \ldots \text{ AND } x_n \text{ is } A_n^k \textbf{ THEN } y \text{ is } B^k \tag{2.1}$$

where N is the number of the rules, $\mathbf{x} = [x_1, x_2, \ldots, x_n]^T$ and y are linguistic variables, A_i^k and B^k, for $i = 1, 2, \ldots, n$, are fuzzy sets, corresponding to input and output of the fuzzy system, respectively; considered in the spaces of real numbers (R^n and R).

In order to know rules (2.1), fuzzy sets A_i^k and B^k, for $i = 1, 2, \ldots, n$ and $k = 1, 2, \ldots, N$, have to be defined. This can be done by an expert or determined by a learning procedure. Usually, Gaussian, triangular or trapezoidal shape membership functions of the fuzzy sets are applied. However, it is not very easy to adjust the proper parameters of these functions. Therefore, learning algorithms are helpful when data are available.

The Gaussian membership functions are expressed as follows:

$$\mu_{A_i^k}(x_i) = \exp\left[-\left(\frac{x_i - \overline{x}_i^k}{\sigma_i^k}\right)^2\right] \tag{2.2}$$

and

$$\mu_{B^k}(y) = \exp\left[-\left(\frac{y - \overline{y}^k}{\sigma^k}\right)^2\right] \tag{2.3}$$

where $\overline{x}_i{}^k$, σ_i^k, $\overline{y}^k$, and σ^k are parameters that can be adjusted during a learning procedure.

In the next chapter, novel explainable recommenders based on neuro-fuzzy systems are proposed, and the learning procedures are applied to fuzzy IF-THEN rules generated by means of the Wang-Mendel (WM) method [53]. Therefore, the WM algorithm for rule generation from data is described in Sect. 2.4.

These learning procedures are employed in neuro-fuzzy recommenders presented in Chap. 3. Similar learning recursions for triangular membership functions can be found e.g. in [40, 52].

In the case when fuzzy IF-THEN rules are formulated by experts, the learning (for tuning parameters) are not necessary. For example, the recommender proposed in Chap. 4 does not need the learning procedures.

Now, we present the learning algorithm based on the steepest descent optimization method to minimize the following objective function [33]:

$$E = \frac{1}{2}(\overline{y} - y^{\star})^2 \tag{2.4}$$

where $y^{\star}$ is the desired real-valued output, and $\overline{y}$ is the output of the system for a particular input value.

For rules (2.1) and membership functions (2.2) and (2.3), the iterative learning method is expressed by recursions [40]:

$$\overline{x}_i{}^k(t+1) = \overline{x}_i{}^k(t) - \eta \frac{\tau_k(\overline{y} - y^{\star})(\overline{y}^k(t) - \overline{y})}{\sum_{j=1}^{N} \tau_j} \frac{2(x_i^{\star} - \overline{x}_i{}^k)}{(\sigma_i^k(t))^2} \tag{2.5}$$

$$\sigma_i{}^k(t+1) = \sigma_i{}^k(t) - \eta \frac{\tau_k(\overline{y} - y^{\star})(\overline{y}^k(t) - \overline{y})}{\sum_{j=1}^{N} \tau_j} \frac{2(x_i^{\star} - \overline{x}_i{}^k)^2}{(\sigma_i^k(t))^3} \tag{2.6}$$

$$\overline{y}^k(t+1) = \overline{y}^k(t) - \eta \frac{\tau_k(\overline{y} - y^{\star})}{\sum_{j=1}^{N} \tau_j} \tag{2.7}$$

where $\overline{x}_i{}^k$, σ_i^k, $\overline{y}^k$, are parameters of membership functions (2.2) and (2.3), centers and widths, respectively, for $i = 1, 2, \ldots, n$ and $k = 1, 2, \ldots, N$, adjusted by the iterative learning procedure, for $t = 0, 1, 2 \ldots$; the constant η denotes the stepsize of the steepest descent algorithm, τ_k is the rule firing level given by formula (2.8).

Let us notice that the width parameters of membership functions (2.3) are not included in recursions (2.5), (2.6), and (2.7). With regard to fuzzy sets B^k, only centers of the membership functions, for $k = 1, 2, \ldots, N$, are significant. This is

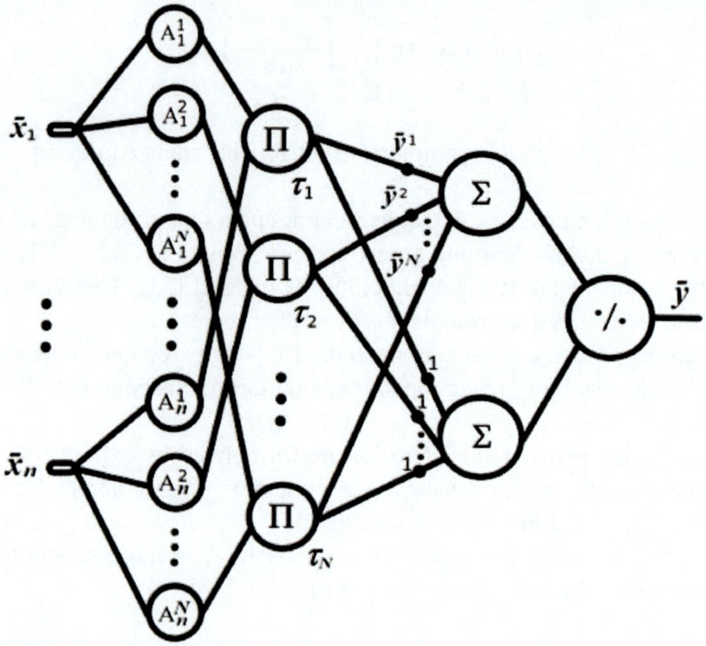

Fig. 2.1 Interpretable neuro-fuzzy network

illustrated in the connectionist architecture of the neuro-fuzzy system presented in Fig. 2.1, in the next subsection.

2.3 Interpretability and Explainability of Neuro-Fuzzy Recommenders

The difference between interpretability and explainability is described in Sect. 1.4. Interpretability refers to the relationship concerning input and output data. Machine learning methods discover this kind of relationship in datasets, however, with different degrees of interpretability (vs. accuracy)—see Fig. 1.5.

Explainability requires interpretability. This means that without interpretability, the explanation is not possible. For interpretable models, various forms of explanation can be applied.

Fuzzy systems, with their inference based on the rules of the form (2.1), formulated by experts, are fully interpretable models. In addition, an explanation can be provided by the use of input data and the rules. In this case, we assume that the rules are known and semantically understandable.

As mentioned in Sect. 2.2, if the rules are not known, they can be generated from examples included in data, and parameters of the fuzzy sets in the rules can be adjusted by a learning algorithm. As a matter of fact, such a system is a combination of a fuzzy system and a neural network, and it is a kind of neuro-fuzzy systems.

In this book, a connectionist form of the neuro-fuzzy system, also called the "neuro-fuzzy network", is considered. The architecture (structure) of the system is similar to multilayer neural networks (MLPs—multilayer perceptrons); see e.g. [59]. However, unlike neural networks, the neuro-fuzzy system is not a "black box". This is an interpretable model because the connectionist architecture reflects the rules, and the mathematical formula that describe the relationship between input and output of the system is understandable. The details are explained below. This means that unlike the mathematical models of neural networks that contain neuron's weights, the neuro-fuzzy system is described by an equation that includes membership functions of fuzzy sets in the rules or their parameters. The connectionist architecture (neuro-fuzzy network) is presented in Fig. 2.1.

As a matter of fact, this is a one example of different neuro-fuzzy architectures; for details see [40]. Other neuro-fuzzy structures are constructed for various kinds of fuzzification, defuzzification, and aggregation operations employed in a fuzzy system.

Crisp numerical values at the input and output of the neuro-fuzzy network are $\overline{\mathbf{x}} = [\overline{x}_1, \overline{x}_2, \ldots, \overline{x}_n]^T$ and $\overline{y}$„ respectively. First layer of the network includes elements that realize membership functions of fuzzy sets A_i^k, for $i = 1, 2, \ldots, n$ and $k = 1, 2, \ldots, N$, in rules (2.1). Thus, at the outputs of these elements we obtain $\mu_{A_i^k}(\overline{x}_1), \ldots, \mu_{A_i^k}(\overline{x}_n)$. Next layer contains elements of product operation, with τ_k at their outputs—representing the rule firing levels (rule activation levels) expressed by the following formula:

$$\tau_k = \Pi_{i=1}^{n} \mu_{A_k^i}(\overline{x}_i) \tag{2.8}$$

Apart from (2.8), other expressions can be used as the rule firing level, for example the min operation is often applied; for details see e.g. [40, 41].

The next part of the network depicted in Fig. 2.1 refers to the conclusion (THEN y is B^k) of rules (2.1). Values $\overline{y}^k$ are centers of the Gaussian membership functions of fuzzy sets B^k, denoted as $\mu_{B^k}(y)$, for $k = 1, 2, \ldots, N$. This part of the neuro-fuzzy architecture refers to the defuzzification method that is determined from the discrete version of the COA (center of area) and called the center average (CA) defuzzification; see e.g. [40].

The mathematical formula that describes the neuro-fuzzy system presented in the connectionist form in Fig. 2.1, is expressed as follows:

$$\overline{y} = \frac{\sum_{k=1}^{N} \tau_k \, \overline{y}^k}{\sum_{k=1}^{N} \tau_k} \tag{2.9}$$

where τ_k is given by (2.8). This is the basic neuro-fuzzy architecture with the assumption that the center-average defuzzification is applied; for details see [40]. In addition, we assume that $\overline{\mathbf{x}} = [\overline{x}_1, \overline{x}_2, \ldots, \overline{x}_n]^T$ and $\overline{y}$ are crisp (not fuzzy) values in spaces of real numbers.

The neuro-fuzzy network presented in Fig. 2.1, and described by mathematical formulas (2.9) with (2.8), reflects the fuzzy IF-THEN rules (2.1). Thus, we see that the system is interpretable, unlike a neural network.

Figure 2.2 portrays a shorter version of the neuro-fuzzy network that is sufficient for classification, and even more interpretable. It is straightforward to explain system performance based on this model.

The neuro-fuzzy network illustrated in Fig. 2.1 is more general, and can be employed for regression problems (including recommendations), as well as for control; it also works for classification but in this case the simple system, shown in Fig. 2.2, is more convenient (if we know rules).

The network shown in Fig. 2.1 is easier to learn by use of the algorithm expressed by Eqs. (2.5), (2.6), and (2.7). This method is similar to the backpropagation applied in neural networks; for detail see e.g. [40]. For the Gaussian membership functions, given by formulas (2.2) and (2.3), parameters $\overline{x}_i{}^k$, σ_i^k, $\overline{y}^k$ are adjusted in this way; as mentioned earlier, σ^k can be ignored.

Recommender systems can be realized by performing a regression task (when real values at the output produce a ranking of recommendations) or as a classifier (e.g. with two classes—recommend and not recommend). The former is suitable for the neuro-fuzzy network shown in Fig. 2.1, while the latter can use the connectionist network presented in Fig. 2.2. The rules for the neuro-fuzzy recommender shown in Fig. 2.2 can be obtained from experts or generated from data. The parameters of the fuzzy sets can be adjusted by the learning algorithm (similar to the backpropagation) applied to the network portrayed in Fig. 2.1. The learning procedure is based on the error (difference between $\overline{y}$ and the desired output for the input $\overline{\mathbf{x}} = [\overline{x}_1, \overline{x}_2, \ldots, \overline{x}_n]^T$. The MSE (Mean Square Error) is often used and minimized during the learning from data.

As mentioned earlier, when we know the fuzzy IF-THEN rules, we can easily explain the performance of the system depicted in Fig. 2.2. The output of the system shows the rule activation levels, so the rule with the maximal firing level should be chosen. The decision (recommendation) indicated in conclusion (THEN part) of this rule is inferred because the input matches this rule with the highest degree.

The neuro-fuzzy systems illustrated in Figs. 2.1 and 2.2 are basic connectionist structures. Many other neuro-fuzzy architectures can be used; for details, see e.g. [40, 41]. For the systems presented in this book, an approach that allows solving the problem of compromise between the system error (accuracy) and the number of optimized parameters describing the system is applied. Of course, there is a trade-off between the system complexity, accuracy, interpretability, and explainability.

The important reason for combining neural networks with fuzzy systems is the learning capability. However, tuning the weights and parameters of fuzzy IF-THEN rules by a learning procedure destroy the semantics of a fuzzy system. But the combination in the form of a neuro-fuzzy system is also useful from the interpretability and explainability point of view, even in the case when rule are known from experts and the learning is not employed. Such a neuro-fuzzy system is proposed in Chap. 4; see also Chap. 5. As a matter of fact, this neuro-fuzzy architecture, first of all represents rules of the fuzzy system. Thus, it can be used for both learning and explanation.

Fig. 2.2 Interpretable neuro-fuzzy network for classification

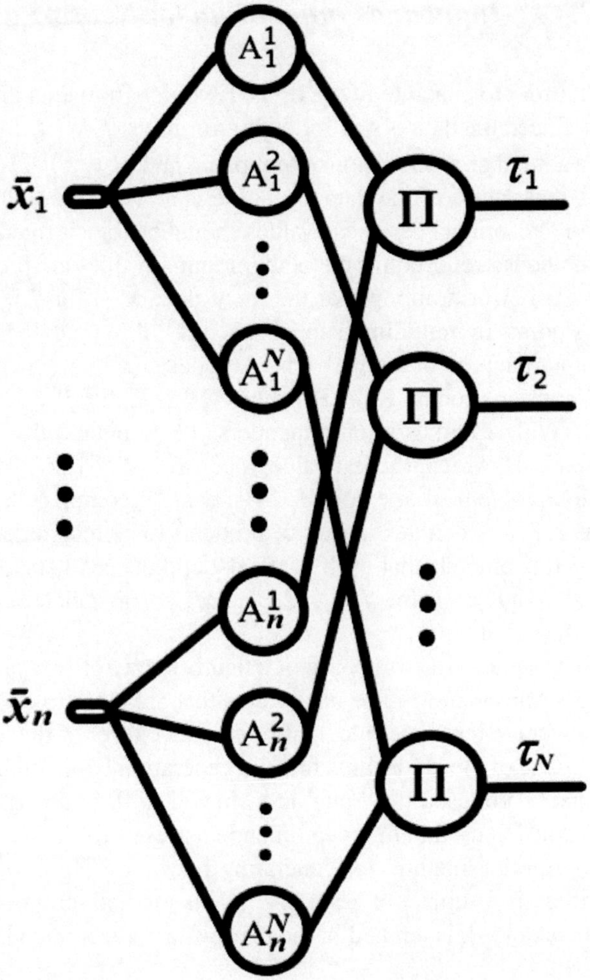

2.4 Rule Generation From Data

The methods for rule generation, applied in the neuro-fuzzy recommenders described in the next chapters, are presented in this section. These methods allow generating fuzzy IF-THEN rules from data. They are instrumental in the case when we do not have rules formulated by experts. However, it is also possible to combine the rules generated from data with expert rules. With regard to recommenders, we can use both datasets and expert knowledge. Before the specific algorithms, the data should be considered.

2.4.1 Input and Output Data for Neuro-Fuzzy Recommenders

In order to generate fuzzy IF-THEN rules from data or learn neuro-fuzzy networks, we need the data of the following form: $(\mathbf{x}^{(j)}; y^{(j)})$, for $j = 1, 2, \ldots, M$, where M is a number of the input-output pairs, and $\mathbf{x}^j = x_1^{(j)}, x_2^{(j)}, \ldots, x_n^{(j)}$. The assumption is that values of the data are numerical. Therefore, when nominal data are applied for a recommender, those values should be transformed into numerical ones. Special methods proposed for encoding nominal values are presented in Chap. 3.

It is worth noting that the input data, $\mathbf{x}^{(j)}$, for $j = 1, 2, \ldots, M$, can be viewed as points in multidimensional space $\mathbf{R}^n$ of real numbers. Thus, $\mathbf{x}^{(j)} \in \mathbf{R}^n$, when the input data are numerical. The data pairs $(\mathbf{x}^{(j)}; y^{(j)})$ are considered in the space of Cartesian product $\mathbf{R}^n \times \mathbf{R}$, where $y^{(j)} \in \mathbf{R}$, for $j = 1, 2, \ldots, M$.

With regard to recommmenders, the output usually refers to the one-dimensional space, $\mathbf{R}$, with particular values, such as $0, 1$ or $1, 2, 3, 4, 5$, typical for recommendation problems. For example, 1—means "recommend", 0—"not recommend", while $1, 2, 3, 4, 5$ denotes ratings of products to be recommended or not recommended.

It is obvious that each of the data points can generate one IF-THEN rule, where ($\mathbf{x}^{(j)}$ and $y^{(j)}$), for $j = 1, 2, \ldots, M$, are associated with the IF and THEN part, respectively.

Of course, the rule base of a neuro-fuzzy (or fuzzy) system should contain much less rules than M. The data points that are close to each other in the attribute space constitute the same rule, that refers to a cluster of the data points.

One of the algorithms for rule generation based on the data points in the attribute space, proposed by Wang and Mendel [53], is described in the next section. This method is used in the recommenders presented in this book.

Another method for generating fuzzy IF-THEN rules, similar to the WM algorithm, is a subject of Sect. 2.4.3. This method, proposed by Nozaki, Ishibushi, and Tanaka [34], is applied in the neuro-fuzzy recommender considered in Chap. 3.

2.4.2 Wang-Mendel Method of Rule Generation

This method has been introduced by Wang and Mendel [53] in 1992, and applied later—also with some modifications; see e.g. [2, 54].

The first step of the original Wang-Mendel (WM) algorithm is to divide the input and output spaces into fuzzy regions.

Let $[x_1^-, x_1^+]$, $[x_2^-, x_2^+]$, ..., $[x_n^-, x_n^+]$, and $[y^-, y^+]$ are domain intervals of input and output data, respectively. The domain intervals are divided into several regions, and membership functions of fuzzy sets are assigned to these regions, as Fig. 2.3 illustrates; see also Fig. 2.4.

In this case, the domain intervals are divided into 5 regions; their length can be equal or unequal. The triangular membership functions, as shown in this figure, can be applied, as well as e.g. Gaussian—defined by (2.2) and (2.3). Different parameters

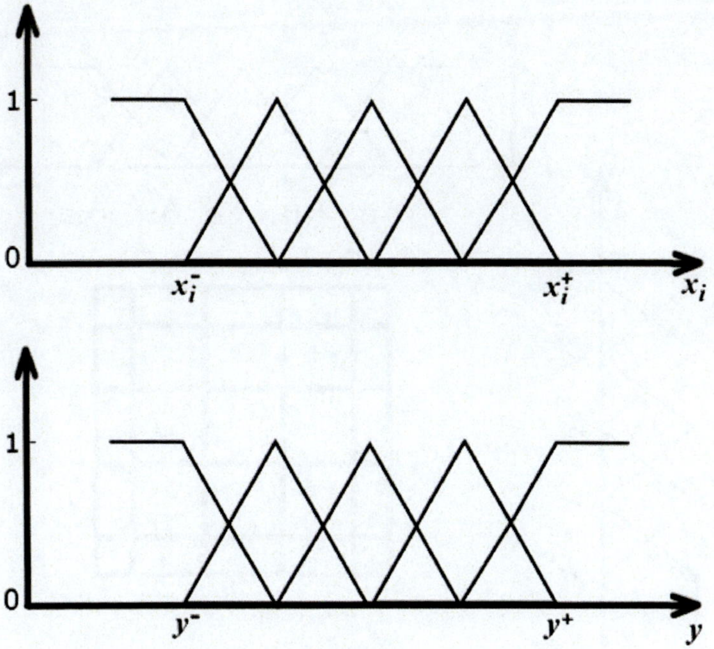

Fig. 2.3 Membership functions within domain intervals

of the membership functions (centers and widths) determine semantic meaning of the fuzzy sets; for example VL—Very Low, L—Low, M—Medium, B—Big, VB—Very Big. The semantic meaning can refer to $A_{11}, A_{12}, \ldots, A_{15}$ and $A_{21}, A_{22}, \ldots, A_{25}$, respectively, in Fig. 2.4.

Instead of 5 regions, it is often sufficient to divide a domain interval into 3 regions, with fuzzy sets: L—Low, M—Medium, B—Big. It is also possible to divide a domain interval into seven regions or more and assign appropriate semantic labels. Of course, the number of the fuzzy sets depends on the data and can be different for particular domains of x_i, y, for $i = 1, 2, \ldots, n$.

The next (second) step of the WM algorithm is to generate fuzzy rules from given data pairs. At first, the degrees of $(\mathbf{x}^{(j)}; y^{(j)})$, for $j = 1, 2, \ldots, M$, in different regions are determined by use of the membership functions (see Fig. 2.3). With regard to fuzzy sets $A_{i1}, A_{i2}, \ldots, A_{i5}$, shown in Fig. 2.4, for $i = 1, 2$, values of membership functions $\mu_{A_{i1}}(x_i^{(j)}), \ldots, \mu_{A_{i5}}(x_i^{(j)})$, and $\mu_{B_1}(y^{(j)}), \ldots, \mu_{B_5}(y^{(j)})$, are calculated. Then, the particular data pairs are assigned to the regions with the maximum degree (the region assigned to the membership function with maximal value for this data point). Finally, for every data pair, a rule is obtained, where the IF and THEN parts are associated with the input and output data, respectively.

The third step is to assign a degree to each rule. Usually, there are many data pairs, and each of them generates one rule during the second step of this algorithm. Thus, there are probably some conflicting rules—with the same IF part and different THEN part. In order to solve this problem, a degree is assigned to each of the rules obtained

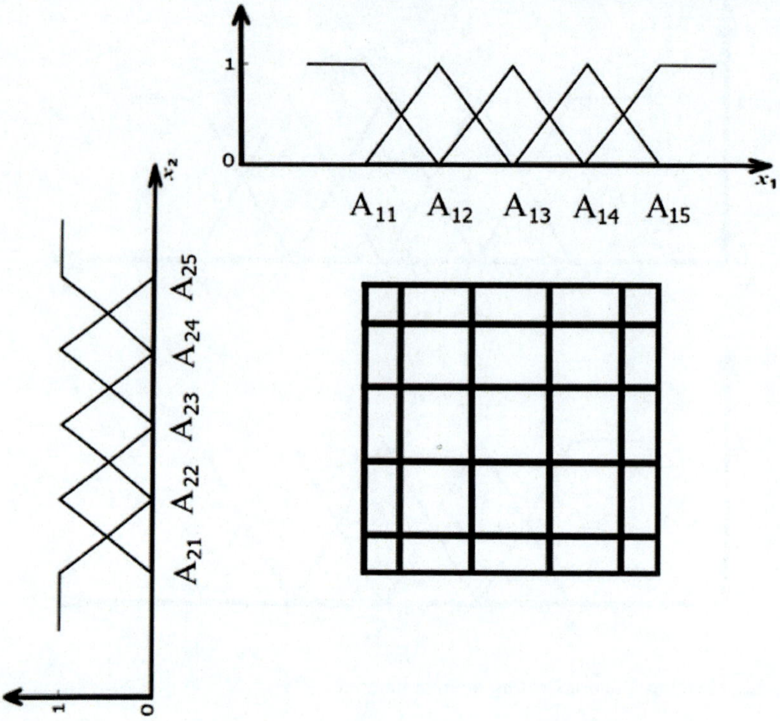

Fig. 2.4 Example of 25 two-dimensional fuzzy regions; for 5 membership functions

from the data pairs. Then, only the rule with the maximum degree is accepted from the conflict group. In this way, the conflict problem is resolved, and also the initial number of rules is greatly reduced.

Now we present more details of this algorithm, with mathematical formulas. Let us assign linguistic values to the membership functions, for the input and output domains, as follows:

$$\begin{array}{r}
x_1 : A_{11}, A_{12}, \ldots, A_{1L_1} \\
x_2 : A_{21}, A_{22}, \ldots, A_{2L_2} \\
\ldots\ldots\ldots\ldots\ldots\ldots\ldots\ldots \\
x_n : A_{n1}, A_{n2}, \ldots, A_{nL_n} \\
y : B_1, B_2, \ldots, B_{L_y}
\end{array} \quad (2.10)$$

As we see, in this case, we consider different numbers of fuzzy sets for particular domains that correspond to linguistic variables in rules (2.1), and can be e.g. of the form presented in Fig. 2.3.

The maximum degrees—during the second step of the WM algorithm—are determined according to the following equations:

$$\mu_{A_i^j}(x_i^{(j)}) = \max_{l=1,2,\ldots,L_i} \{\mu_{A_{il}}(x_i{}^{(j)})\} \tag{2.11}$$

for $i = 1, 2, \ldots, n$, and

$$\mu_{B^j}(y^{(j)}) = \max_{l=1,2,\ldots,L_y} \{\mu_{B_l}(y^{(j)})\} \tag{2.12}$$

In this way M fuzzy IF-THEN rules of the form (2.1) are obtained; for each data pair $(\mathbf{x}^{(j)}; y^{(j)})$, $j = 1, 2, \ldots, M$.

In the third step of the WM algorithm, a degree is assigned to each of the M rules, by use of formulas (2.11) and (2.12), as follows:

$$D(R^j) = \mu_{A_1^j}(x_1^{(j)})\mu_{A_2^j}(x_2^{(j)}) \ldots \mu_{A_n^j}(x_n^{(j)})\mu_{B^j}(y^{(j)}) \tag{2.13}$$

for $j = 1, 2, \ldots, M$.

As a matter of fact, the degree (2.13) is obtained as a product of the value of rule firing level (2.8) and the membership of the consequent fuzzy set; for the data pair that generates this rule.

The degree, $D(R^j)$, assigned to each rule R^j, for $j = 1, 2, \ldots, M$, allows reducing the number of the rules, including into the rule base of a fuzzy system, only the rules with the maximal degree in particular regions. Thus, starting from M rules, generated by M data pairs, the WM algorithm produces the rule base of a fuzzy system composed of the reduced number of N rules of the form (2.1), where $N \leq M$.

Figure 2.4 illustrates the partitioning of the attribute space in case of two attributes. As mentioned earlier, the fuzzy regions are determined by the membership functions of fuzzy sets (2.10).

Of course, different numbers and shapes of the membership functions can be used, and not necessary equally distributed as shown in Figs. 2.3 and 2.4. The membership functions can be determined by experts in a different way. In such a case, the WM algorithm is employed in the recommender proposed in Chap. 4.

2.4.3 Nozaki-Ishibuchi-Tanaka Method

The Nozaki-Ishibuchi-Tanaka (NIT) method [34] is similar to the WM algorithm. This is a method for automatically generating fuzzy IF-THEN rules from numerical data.

However, the simpler form of fuzzy rules (2.1) is considered—with nonfuzzy (crisp) real numbers in the consequent parts:

$$\textbf{IF } x_1 \text{ is } A_1^k \text{ AND } x_2 \text{ is } A_2^k \text{ AND} \ldots \text{ AND } x_n \text{ is } A_n^k \textbf{ THEN } y = b^k \tag{2.14}$$

for $k = 1, 2, \ldots, N$.

The number of the rules, $N = L_1 L_2 \dots L_n$, depends on the number of fuzzy sets (2.10). Figure 2.4 illustrates an example of 5 triangular membership functions for two linguistic variables, x_1 and x_2, that produce $5 \cdot 5 = 25$ fuzzy regions within the two-dimensional domain. The same is considered with regard to the WM algorithm. A single fuzzy IF-THEN rule is generated in each region of the domain space.

This means that fuzzy sets A_i^k in rules (2.14) are replaced by fuzzy sets (2.10), using every combinations of $A_i^k = A_{il_i}$, for $i = 1, 2, \dots, n$ and $l_i = 1, 2, \dots, L_i$. When fuzzy sets portrayed in Fig. 2.4 are applied, the rule:

$$\textbf{IF } x_1 \text{ is } A_1^k \text{ AND } x_2 \text{ is } A_2^k \textbf{ THEN } y = b^k \tag{2.15}$$

can be one of 25 combinations, for example:

$$\textbf{IF } x_1 \text{ is } A_{11} \text{ AND } x_2 \text{ is } A_{25} \textbf{ THEN } y = b^k \tag{2.16}$$

where index k corresponds to the specific rule. This means that rule (2.16) is associated with the region of the left upper corner in Fig. 2.4.

The NIT heuristic method determines the consequent real numbers, b^k, in rules (2.14) as follows:

$$b^k = \frac{\sum_{j=1}^{M} \omega_k \, y^{(j)}}{\sum_{j=1}^{M} \omega_k} \tag{2.17}$$

where

$$\omega_k = [\tau_k(\mathbf{x}^{(j)})]^\alpha \tag{2.18}$$

and τ_k denotes the rule activation levels that can be obtained e.g. by use of equation (2.8), for input data $\mathbf{x}^j = x_1^{(j)}, x_2^{(j)}, \dots, x_n^{(j)}$. Let us notice the constant α that plays a specific role, depending on its value: $0 < \alpha < 1$, $\alpha = 1$, and $\alpha > 1$ that is explained in [34].

Values ω_k, given by (2.18), are viewed as weights of the input-output data pairs $(\mathbf{x}^{(j)}; y^{(j)})$, for $j = 1, 2, \dots, M$. If $\alpha > 1$, much of only the data pairs with high degrees of compatibility with the rule is taken into account. In the case of $0 < \alpha < 1$, much of even data pairs with low degrees of compatibility as well as those with high degrees of compatibility are included.

Let us notice that for $\alpha = 1$, formula (2.17) corresponds to equation (2.9), in the case when $N = M$ and $y^{(j)}$, for $j = 1, 2, \dots, M$, is viewed as the center $\overline{y}^j$ of a fuzzy set B^j in rules (2.1). This refers to the situation when every data pair generates one rule of this form. Let us imagine that every data pair $(\mathbf{x}^{(j)}; y^{(j)})$, for $j = 1, 2, \dots, M$, is assigned to the rule (2.1), where $y^{(j)} = \overline{y}^j$ that is the center of B^j.

However, the number of data pairs is usually greater than the number of rules, $M > N$. Formulas (2.17) and (2.18) allow to determine the reduced number of rules, N, from M data items. Of course, values b^k, for $k = 1, 2, \dots, M$, obtained in this way, can be different than centers, $\overline{y}^k$, of fuzzy sets B^k in rules (2.1). This is illustrated in Figs. 3.2 and 3.3, with regard to the recommender presented in Chap. 3.

The NIT method allows to determine rules with values, in the consequent (THEN) part of the rules, more adjusted to the data. This is significant especially in the case when $y^{(j)}$, for $j = 1, 2, \ldots, M$, are located not close to the centers of fuzzy sets in consequent part of rules (2.1). As a matter of fact, this is important mostly in control applications.

However, it is worth mentioning that with regard to recommendation systems, this method seems to be even more suitable than the WM algorithm because it generates rules with real numbers in their consequent parts. Of course, if the output singleton values are known the rules can be formulated without any learning procedure.

2.5 Fuzzy IF-THEN Rules in Recommendation Problems

The simplified fuzzy rules of the form (2.14) have been often used. It was proven [16] that fuzzy rule-based systems with the simplified rules could approximate any nonlinear function on a compact set to arbitrary accuracy under certain conditions; see also [34].

The simplified rules (2.14) can be viewed as a special case of rules (2.1) as well as a special case of fuzzy rules for the Takagi-Sugeno-Kang (TSK) fuzzy system [47, 48], the so-called "Zero-Order Takagi-Sugeno-Kang" (ZO-TSK).

The fuzzy system with the inference based on rules (2.1) is often called the Mamdani system (see e.g. [28, 29, 40, 41]). The difference between the Mamdani and TSK systems is visible in the consequent parts of the rules (THEN parts). In the TSK rules, instead of the fuzzy sets (B^k), there is a function $y_k = f^{(k)}(x_1, x_2, \ldots, x_n)$, in the general form, for $k = 1, 2, \ldots, N$, that can be e.g. a linear function, and a constant value in the simplest case.

Both systems, in their simplest form, are described by the following mathematical formula:

$$\overline{y} = \frac{\sum_{k=1}^{N} \tau_k \, v_k}{\sum_{k=1}^{N} \tau_k} \tag{2.19}$$

where τ_k is the rule activation level, according to (2.8), and

$$v_k = \begin{cases} \overline{y}^k & \text{for the Mamdani system} \\ b^k & \text{for the ZO-TSK system} \end{cases} \tag{2.20}$$

for $k = 1, 2, \ldots, N$.

As we see in Eq. (2.20), and as explained above, the Mamdani system refers to the rule base (2.1), where $\overline{y}^k$ denotes the centers of the Gaussian (or triangular) membership functions of fuzzy sets B^k, while b^k represents constant values in rules (2.14).

It is obvious that the NIT method, described in Sect. 2.4.3, can be applied to the ZO-TSK system, determining values v_k by Eq. (2.17). As a matter of fact, the same rules can be employed for the Mamdani system with singletons as the consequent fuzzy sets B^k, $k = 1, 2, \ldots, N$, and for the ZO-TSK system.

The fuzzy IF-THEN rules and systems considered in this section are applied in recommenders presented in the next chapters. However, some modifications have been introduced.

The recommendation problem can be described as follows: Let us consider a database of M objects characterized by n attributes (features) taking numerical or nominal values. For every object, along with the attribute values, the rating value provided by users is associated. A recommender should decide if other similar objects should be recommended to particular users according to their preferences. The decision concerning the recommendations is inferred by the system based on the data pairs (attribute values, rating value) included in the dataset.

In case of numerical values of the attributes, the data pairs for the recommendation problem correspond to the input-output data pairs $(\mathbf{x}^{(j)}; y^{(j)})$, for $j = 1, 2, \ldots, M$, considered in the previous sections. Otherwise, methods for data preparation, including encoding nominal values, are proposed.

2.6 Classification in Recommenders

Neuro-fuzzy systems, in general, are mostly employed in control problems, but also in combinations of other areas of applications. Usually, three main kinds of problems are distinguished with regard to neuro-fuzzy systems: control, classification, and approximation (see e.g. [32]). With regard to recommenders, the classification and approximation are considered rather than control. Therefore, this section concerns classifiers.

2.6.1 Neuro-Fuzzy Classifiers

According to [32], advantages of using neuro-fuzzy systems as classifiers are the interpretability of fuzzy classifiers and the possibility of integrating prior knowledge in the form of fuzzy rules.

Although neuro-fuzzy systems, as presented in previous sections of this chapter, can be used in order to solve control tasks, as well as approximation and classification problems, we focus our attention on explainable recommenders. From this point of view, we are interested in fuzzy IF-THEN rules for recommendations.

Comparing the neuro-fuzzy systems shown in Figs. 2.1 and 2.2, we know that both of them can be employed as classifiers but the latter allows for better explanation of the results than the former one. However, the network portrayed in Fig. 2.1 can also be applied as a controller and a function approximator.

Fuzzy IF-THEN rules for recommendations usually have single crisp values in their antecedent parts, or fuzzy numbers (see e.g. [9]). In the former case the rules of the form (2.14) are cosidered while in the latter—rules (2.1) where fuzzy sets in the consequent parts are fuzzy numbers.

In recommenders, classification rules with consequent parts that include one of different values, indicating two classes; for example: 1—recommend, 0—not recommend (or 1—Yes, and 0—No), are applied. Apart from this case, more crisp or fuzzy numbers are employed as values in consequent parts of the rules, e.g. 1, 2, 3, 4, 5, as assessments of recommended (or not recommended) products.

It is worth noticing that a fuzzy number is a special case of a fuzzy set (see e.g. [9]). Thus, it is obvious that—for example—numbers 1, 2, 3, 4, 5 as the assessments can be treated as crisp values or viewed as fuzzy numbers with linguistic labels such as: approximately 1, 2, 3, 4, and 5, respectively.

As presented above, usually the classification concerns more than one class. Such problems are discussed with regard to the recommenders described in Chap. 3. In particular, two-class problems refer to the binary classification. However, we can also apply a one-class classifier as a recommender. A real example of such a recommendation system is proposed in Chap. 4. Therefore, classification systems for one-class problems are described in the next subsection.

Different aspects of fuzzy classifiers are presented in [25]. As mentioned earlier, a neuro-fuzzy system for classification is proposed in [32]. According to authors of this publication: neuro-fuzzy classifiers offer tools that allow to obtain fuzzy classification rules by a learning algorithm. This gives the possibility to get an appropriate fuzzy classifier by learning from data. Although it can be difficult to find, in this way, a classifier that can be easily interpreted, the main reason for using a fuzzy approach for classification is to have an interpretable classifier.

Most of fuzzy and neuro-fuzzy classifiers concern problems of multi-class classification; see e.g. [7, 11].

2.6.2 One-Class Classifiers

Apart from the binary and multi-class classifiers, also one-class classification methods and systems are studied, applied, and presented in publications, e.g. [17, 21, 31, 35, 38, 39, 42, 46, 56]. The goal of the one-class classification is to recognize objects (items) that belong to one class, having learning data only from the one class. Thus, unlike in the case of binary classification, we do not have data from another class [14]. Speaking more precisely, in the one-class classification—as a matter of fact, we consider two classes (usually referred to as the positive and negative, or target and non-target class, respectively). But the target class is well represented by the training data, whereas the non-target class has either no data items at all, or only few that do not sufficiently characterize the second class. For details, see e.g. [20, 49, 50].

There are also publications on fuzzy approaches to one-class classification, see e.g. [5, 13]. Thus, one-class classifiers are considered within classical Machine Learning methods. As mentioned earlier, the recommender presented in Chap. 4 produces recommendations solving a one-class classification problem by use of real data that represent only one class.

The one-class classifier proposed in Chap. 4 is based on fuzzy IF-THEN rules, and it is interpretable and explainable.

2.6.3 Classification in Content-Based Recommender Systems

The goal of a classification is to predict a class to which new input data items belong, based on past observations (usually included in a dataset). As explained in Chap. 1, content-based recommendation systems try to recommend items that are similar to those a given user preferred in the past. Unlike in recommenders that use collaborative filtering, other users, and their preferences, do not play a significant role.

In content-based recommenders, attributes of items are employed in order to produce recommendations. The term "content" refers to the item descriptions (values of the attributes). Collaborative filtering methods do not use item attributes, and attempt to recommend items by identifying other users with similar tastes.

Content-based systems are closely related to knowledge-based recommenders. Such systems incorporate knowledge from content features. Hence, they often provide highly interpretable recommendation process, and offer an explanation. For details, see also [1].

In content-based systems, the item descriptions—as data pairs of attribute values and labels that include ratings provided by users—are applied as learning (training) data in a classification or regression problem. It is worth noticing that the regression can be viewed as an approximation, and in a discrete case—as a classification.

All the neuro-fuzzy recommenders presented in this book are content-based, and knowledge-based recommenders, that perform a classification (or regression) task, and are interpretable and explainable.

Of course, we can observe differences concerning interpretability and accuracy with regard to the proposed neuro-fuzzy recommenders. For example, it is known that Mamdani systems are more interpretable but TSK systems are more accurate and computationally efficient [45, 51]. This is because the IF-THEN rules in the Mamdani systems include fuzzy sets in both the antecedent and consequent parts while in the TSK systems only in the antecedent part. However, this conclusion is more suitable with regard to control tasks than the classification in recommenders. Nevertheless, the more complex a system is the more difficult to interpret and explain. From this point of view, we can compare the recommendation systems described in Chap. 3.

References

1. Aggarwal, C.C.: Recommender Systems. Springer, Berlin (2016)
2. Alvarez-Estevez, D., Moret-Bonillo, V.: Revisiting the Wang-Mendel algorithm for fuzzy classification. Exp. Syst. **35**(4) (2018)
3. Anderson, J.A.: An Introduction to Neural Networks. The MIT Press, London (1995)
4. Bishop, C.M.: Pattern Recognition and Machine Learning. Springer, Berlin (2006)
5. Bosco, G.L., Pinello, L.: A fuzzy one class classifier for multi layer model. Fuzzy Logic Appl. **5571**, 124–31 (2009). LNCS
6. Buchanan, B.G., Shortliffe, E.H.: Rule-Based Expert Systems: The MYCIN Experiments of the Stanford Heuristic Programming Project. Addison-Wesley (1985)
7. Chakraborty, D., Pal, N.: A Neuro-Fuzzy scheme for simultaneous feature selection and fuzzy rule-based classification. IEEE Trans. Neural Netw. **15**(1), 110–123 (2004)
8. Czogała, E., Łeski, J.: Fuzzy and Neuro-Fuzzy Intelligent Systems. Physica-Verlag. A Springer-Verlag Company. Heidelberg, New York (2000)
9. Dubois, D., Prade, H.: Fuzzy Sets and Systems: Theory and Applications. Academic, London (1980)
10. Duning, T., Friedman, E.: Practical Machine Learning: Innovations in Recommendation. O'Reilly Media, Inc. (2014)
11. Ghosh, A., Shankar, B.U., Meher, S.K.: A novel approach to neuro-fuzzy classification. Neural Netw. **22**, 100–109 (2009)
12. Grossberg, S.: Neural Networks and Artificial Intelligence. The MIT Press, Cambridge, MA (1988)
13. Hao, P.Y.: Fuzzy one-class support vector machines. Fuzzy Sets Syst. **159**, 2317–2336 (2008)
14. He, H., Ma, Y.: Imbalance Learning: Foundations, Algorithms, and Applications. Wiley-IEEE Press (2013)
15. Jackson, P.: Introduction to Expert Systems, 3rd edn. Addison-Wesley Longman Publishing Co., Inc, Boston, MA, USA (1998)
16. Jang, J.S.R.: ANFIS: adaptive-network-based fuzzy inference system. IEEE Trans. Syst. Man Cybern. **23**(3), 665–685 (1993)
17. Juszczak, P., Tax, D.M.J., Pekalska, E., Duin, R.P.W.: Minimum spanning tree based one-class classifier. Neurocomputing **72**(7–9), 1859–1869 (2009)
18. Kasabov, N.K.: Foundations of Neural Networks, Fuzzy Systems, and Knowledge Engineering. The MIT Press, Cambridge, MA (1996)
19. Keller, J.M., Hunt, D.: Incorporating fuzzy membership functions into the perceptron algorithm. IEEE Trans. Pattern Anal. Mach. Intell. **7**, 693–699 (1985)
20. Khan, S.S., Madden, M.G.: A survey of recent trends in one class classification. In: Coyle, L., Freyne, J. (eds.) AICS 2009, LNAI 6206, pp. 188–197. Springer, Berlin (2010)
21. Khan, S.S., Madden, M.G.: One-class classification taxonomy of study and review of techniques. Knowl. Eng. Rev. **29**(3), 345–374 (2014)
22. Klir, G.J., Yuan, B. (eds.): Fuzzy Sets, Fuzzy Logic and Fuzzy Sytems: Selected Papers by Lotfi A. Zadeh. Adv. Fuzzy Syst. Appl. Theory **6** (1996)
23. Kosko, B.: Neural Networks and Fuzzy Systems. A Dynamical Systems Approach to Machine Intelligence. Prentice Hall, Englewood Cliffs, New Jersey (1992)
24. Kruse, R., Gebhardt, J., Klawonn, F.: Foundations of Fuzzy Systems. Wiley, New York (1994)
25. Kuncheva, L.: Fuzzy Classifier Design. Studies in Fuzziness and Soft Computing. Springer, Berlin (2000)
26. Lin, C.T.: Neural Fuzzy Control Systems with Structure and Parameter Learning. World Scientific, Singapore (1994)
27. Lin, C.T., Lee, G.C.S.: Neural Fuzzy Systems. Prentice Hall, A Neuro-Fuzzy Synergism to Intelligent Systems (1996)
28. Mamdani, E.H.: Applications of fuzzy algorithm for simple dynamic plant. Proc. IEE **121**(12), 1585–1588 (1974)

29. Mamdani, E.H., Assilian, S.: An experiment in linguistic synthesis with a fuzzy logic controller. Int. J. Mach. Stud. **7**, 1–13 (1975)
30. Medsker, L.R.: Hybrid Neural Network and Expert Systems. Kluwer Academic Publisher (1994)
31. Moya, M., Koch, M., Hostetler, L.: One-class classifier networks for target recognition applications. In : Proceedings of the World Congress on Neural Networks. Portland, OR. International Neural Network Society, INNS. pp. 797–801 (1993)
32. Nauck, D., Klawonn, F., Kruse, R.: Foundations of Neuro-Fuzzy Systems. Wiley, New York (1997)
33. Nomura, H., Hayashi, I., Wakami, N.: A self-tuning method of fuzzy control by descent method. In: Proceedings of the 4th International Fuzzy Systems Association World Congress (IFSA'91). Brussels, Belgium, pp. 155–158 (1991)
34. Nozaki, K., Ishibuchi, H., Tanaka, H.: A simple but powerful heuristic method for generating fuzzy rules from numerical data. Fuzzy Sets Syst. **86**(3), 251–270 (1997)
35. Oza, P., Patel, V.M.: One-class convolutional neural network. IEEE Signal Proces. Lett. **26**(2), 277–281 (2019)
36. Patterson, D.W.: Artificial Neural Networks: Theory and Applications. Prentice Hall (1996)
37. Pedrycz, W.: Fuzzy Control and Fuzzy Systems. Wiley, New York, NY, USA (1993)
38. Perera, P., Patel, V.M.: Learning deep features for one-class classification. IEEE Trans. Image Proces. **28**(11), 5450–5463 (2019)
39. Rätsch, G., Mika, S., Schölkopf, B., Müller, K.-R.: Constructing boosting algorithms from SVMs: an application to one-class classification. IEEE Trans. Pattern Anal. Mach. Intell. **24**(9), 1184–1199 (2002)
40. Rutkowska, D.: Neuro-Fuzzy Architectures and Hybrid Learning. Physica-Verlag. Springer, Heidelberg, New York (2002)
41. Rutkowski, L.: Flexible Neuro-Fuzzy Systems: Structures. Kluwer Academic Publisher, Learning and Performance Evaluation (2004)
42. Sabokrou, M., Khalooei, M., Fathy, M., Adeli, E.: Adversarially learned one-class classifier for novelty detection. In: Proceedings of the IEEE Conference on Computer Vision and Pattern Recognition, pp. 3379–3388 (2018)
43. Sahin, S., Tolun, M.R., Hassanpour, R.: Hybrid expert systems: a survey of current approaches and applications. Exp. Syst. Appl. **39**(4), 4609–4617 (2012)
44. Shihabudheen, K.V., Pillai, G.N.: Recent advances in neuro-fuzzy system: survey. Knowl.-Based Syst. **127**, 100–113 (2017)
45. Shihabudheen, K.V., Pillai, G.N.: Regularized extreme learning adaptive neuro-fuzzy algorthm for regression and classification. Knowl.-Based Syst. **152**, 136–162 (2018)
46. Shin, H.J., Eom, D.W., Kim, S.S.: One-class support vector machines: an application in machine fault detection and classification. Comput. Indus. Eng. **48**(2), 395–408 (2005)
47. Sugeno, M., Kang, G.: Structure identification of fuzzy model. Fuzzy Sets Syst. **28**(1), 15–33 (1988)
48. Takagi, T., Sugeno, M.: Fuzzy identification of systems and its applications to modeling and control. IEEE Trans. Syst. Man Cybern. **15**(1), 116–132 (1985)
49. Tax, D.M.J.: One Class Classification: Concept-Learning in the Absence of Counter-Examples, Ph.D. Thesis. Delft University of Technology (2001)
50. Tax, D.M.J., Duin, R.P.W.: Uniform object generation for optimizing one-class classifiers. J. Mach. Learn. Res. **2**, 155–173 (2001)
51. Tiruneh, G.G., Fayek, A.R., Sumati, V.: Neuro-fuzzy systems in constriction engineering and management research. Automat. Construct. **119** (2020)
52. Wang, L.-X.: Adaptive Fuzzy Systems and Control. PTR Prentice Hall, Englewood Cliffs, New Jersey (1994)
53. Wang, L.-X., Mendel, J.M.: Generating fuzzy rules by learning from examples. IEEE Trans. Syst. Man Cybern. **22**(6), 1414–1427 (1992)
54. Wang, L.-X.: The WM method completed: a flexible fuzzy system approach to data mining. IEEE Trans. Fuzzy Syst. **11**(6), 768–782 (2003)

55. Witten, I., Frank, E., Hall, M., Pal, C.: Data Mining: Practical Machine Learning Tools and Techniques, 4th edn. Morgan Kaufmann (2016)
56. Vert, R., Vert, J.-P.: Consistency and convergence rates of one-class SVMs and related algorithms. J. Mach. Learn. Res. **7**, 817–854 (2006)
57. Yager, R.R., Zadeh, L.A. (eds.): Fuzzy Sets, Neural Networks, and Soft Computing. Van Nostrand Reinhold, New York (1994)
58. Zadeh, L.A.: Towards a theory of fuzzy systems. In: Kalman R.E. and deClaris N. (eds.). Aspect of Network and System Theory. Holt, Rinehart and Winston. New York (1971)
59. Żurada, J.M.: Introduction to Artificial Neural Systems. West Publishing Company (1992)

Chapter 3
Novel Explainable Recommenders Based on Neuro-Fuzzy Systems

Check for updates

3.1 Recommender A

As it was indicated in Chap. 2, it is possible to apply rule-based recommender systems to achieve explainability without losing too much accuracy. Such a system is usually based on fuzzy logic. As rules are, by definition, interpretable by humans, they can be used to generate explanations. However, it is not trivial to generate rules from examples, reduce them, and optimize.

In this section, a new method derived from the WM [31] and NIT [18] algorithms, presented in Sects. 2.4.2 and 2.4.3, respectively, combined with the ZO-TSK fuzzy system (see Sect. 2.5), is considered. It allows dealing with singleton outputs that can be further optimized by the use of the "Wolf Grey Optimizer (GWO)" [8, 16]. It is shown in the literature [5, 9] that such a combination of a neuro-fuzzy system with the GWO improves the performance of the conventional systems.

All experiments are based on the MovieLens 10M dataset [11], and tested on six neuro-fuzzy recommenders (see Sects. 3.1.2 and 3.1.3), denoted as WM-T, NIT-T, NIT-S, WM-T+S, NIT-T+S, NIT-S+S. In the next subsection, a novel method for transforming nominal values into a numerical form is proposed. This allows representing nominal values, e.g. movie genres or actors, in a neuro-fuzzy recommender.

3.1.1 Feature Encoding

As explained earlier, especially in Chap. 2, Sect. 2.5, objects (items) to be recommended are described by attributes (features) that can be of a different type, not necessarily numerical (crisp values). Moreover, multiple nominal values can be assigned to a single item attribute. To process such data by a neuro-fuzzy recommender, an aggregation of nominal values is proposed in this section.

T. Rutkowski, *Explainable Artificial Intelligence Based on Neuro-Fuzzy Modeling with Applications in Finance*, Studies in Computational Intelligence 964,
https://doi.org/10.1007/978-3-030-75521-8_3

For each user and each attribute of items rated by users, a list of unique nominal values is created. The user preferences of each value are determined as an average rate of the item that contains a particular attribute value. Then, input values for the recommender system are calculated as an average preference of all values that occur for a given attribute of an item. This idea is illustrated in Fig. 3.1 on a simple example. In the sequel, it will be extended to a general case. It requires to introduce several formal descriptions.

Denote by $\mathbf{a} = [a_1, a_2, \ldots, a_n]^T$ a general concept of a vector of attributes that describe objects belonging to an object space, $\mathbf{O}$, that contains a set of the objects, e.g. book, movies. Let $\mathbf{a}^{(j)} = [a_1{}^{(j)}, a_2{}^{(j)}, \ldots, a_n{}^{(j)}]^T$ denotes a vector of attributes describing a particular object $\mathbf{o}_j$ that belongs to $\mathbf{O}$. For every object $\mathbf{o}_j$, values of $\mathbf{a}^{(j)}$, along with $d^{(j)}$ that is the decision attribute—label denoting e.g. rating with regard to the movie data, for $j = 1, 2, \ldots, M$, are included in a dataset. Thus, $(\mathbf{a}^{(j)}; d^{(j)})$ are input-output data pairs that can be taken from the dataset, and process by a recommender.

However, the nominal values cannot be introduced directly to the neuro-fuzzy recommender. Therefore, the approach presented in Fig. 3.1 is applied. As we see, some attributes—like "genre"—can have more than one value, e.g. $\{drama, comedy\}$,. Speaking more precisely, the attribute values can be viewed as vectors of the nominal values. Thus, each attribute $a_i{}^{(j)}$, for $i = 1, 2, \ldots, n$ and $j = 1, 2, \ldots, M$, takes a value of the form $\overline{\mathbf{a}}_i = [\overline{a}_{i,j_1}, \overline{a}_{i,j_2}, \ldots, \overline{a}_{i,j_{L_i}}]^T$ where $(j_1, j_2, \ldots, j_{L_i})$ is a subsequence of $(1, 2, \ldots, L_i)$, and L_i denotes the number of the possible nominal values of attribute a_i. In this notation, we assume that there is no difference e.g between $\{drama, comedy\}$ and $\{comedy, drama\}$. It is worth adding that we can also consider the recommendation problem, with an information that a movie is e.g. more comedy than drama or opposite; moreover certain degrees can be assigned to the attribute values.

With regard to the neuro-fuzzy recommenders, the vectors of attribute values, $\overline{\mathbf{a}}_i = [\overline{a}_{i,j_1}, \overline{a}_{i,j_2}, \ldots, \overline{a}_{i,j_{L_i}}]^T$, for $i = 1, 2, \ldots, n$, can refer to the fuzzy sets as linguistic values; see (2.10).

The idea of the feature encoding, illustrated in Fig. 3.1, allows to transform the vectors of nominal attribute values into single numerical values that can be directly process by the neuro-fuzzy recommenders. In this way, we realize a transformation of $(\mathbf{a}^{(j)}; d^{(j)})$ to $(\overline{\mathbf{x}} = [\overline{x}_1, \overline{x}_2, \ldots, \overline{x}_n]^T; \overline{y})$; see Chap. 2, Sect. 2.3. As a matter of fact, we transform one attribute (with multiple nominal values) to another attribute (with single numerical value). In the example shown in Fig. 3.1, the "genre" attribute is replaced by the "genre preference".

3.1.2 Description of the Proposed Recommender A

As the recommender A, the neuro-fuzzy system with the inference based on the simplified rules of the form (2.14) is proposed. As explained in Sect. 2.5, the Mamdani system with singletons in the consequent part of the rules, and the ZO-TSK that uses

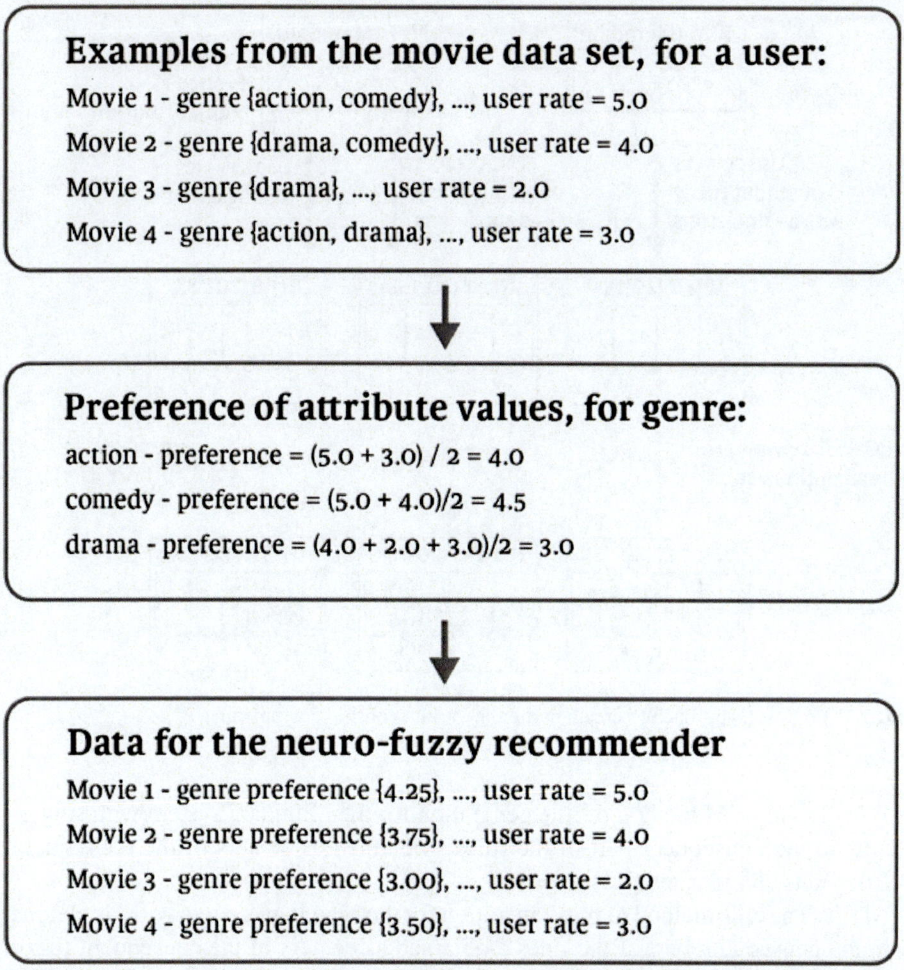

Fig. 3.1 Example of the movie data preparation for the neuro-fuzzy recommender

constant consequent values, employ the simplified rule base that can be rewritten as follows:

$$\textbf{IF } x_1 \text{ is } A_1^k \text{ AND } x_2 \text{ is } A_2^k \text{ AND} \ldots \text{ AND } x_n \text{ is } A_n^k \textbf{ THEN } y = v_k \tag{3.1}$$

for $k = 1, 2, \ldots, N$.

The WM algorithm presented in Sect. 2.4.2, and the NIT method described in Sect. 2.4.3, are employed for rule generation of the Mamdani and ZO-TSK systems, respectively.

The following combinations are considered as the recommender A:

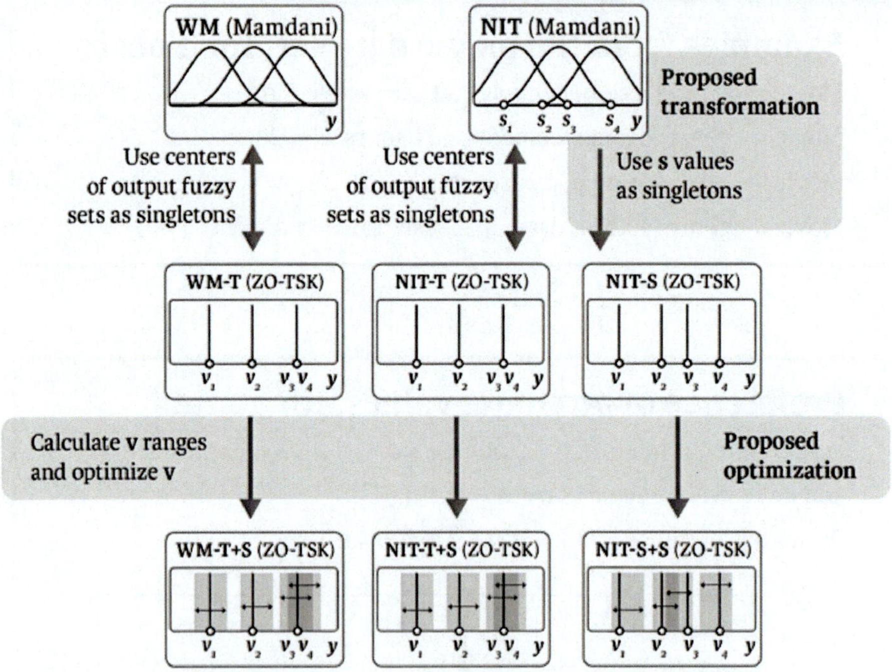

Fig. 3.2 Proposed approach to create particular variants of the recommender A

- WM-T: The WM method of rule generation for the Mamdani system with singletons in the consequent part of the rules—obtained as centers of the consequent fuzzy sets, B^k in rules (2.1), $k = 1, 2, \ldots, N$.
- NIT-T: The NIT method of rule generation for the Mamdani system with singletons in the consequent part of the rules—obtained as centers of the consequent fuzzy sets, B^k in rules (2.1), $k = 1, 2, \ldots, N$.
- NIT-S: The NIT method of rule generation with the consequent constants determined according to Eqs. (2.17) and (2.18), for $\alpha = 1$.
- WM-T+S: The WM-T system optimized by the GWO (Grey Wolf Optimizer).
- NIT-T+S: The NIT-T system optimized by the GWO.
- NIT-S+S: The NIT-S system optimized by the GWO.

As a matter of fact, each of these variants of the recommender A can be viewed as the ZO-TSK system that is equivalent to the Mamdani system with the consequent singletons. Thus, we can say that this kind of recommenders is realized as the ZO-TSK neuro-fuzzy system. The particular variants differ mainly in their rule bases, especially the consequent part of the rules, generated in a different way, and also optimized or not. Figure 3.2 illustrates the proposed approach to create particular systems.

Figure 3.3 portrays the difference between the centers of the consequent fuzzy sets and the consequent constant determined according to the NIT method; for Gaussian membership functions.

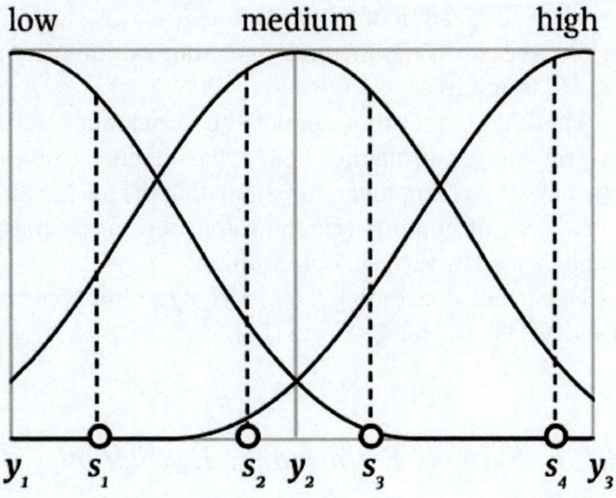

Fig. 3.3 Centers of the fuzzy sets, y_1, y_2, y_3, and constant values from NIT, s_1, s_2, s_3, s_4

In Figs. 3.2 and 3.3, values s_1, s_2, s_3, s_4 correspond to the consequent constants b^k in rules (2.14); see (2.17) and (2.18), while values y_1, y_2, y_3 refer to rules (2.1) where fuzzy sets B^k are singletons (centers of the fuzzy sets, $\overline{y}^1$, $\overline{y}^2$, $\overline{y}^3$, respectively).

Values v_1, v_2, v_3, v_4, in Fig. 3.2, correspond to y_1, y_2, y_3 or s_1, s_2, s_3, s_4, respectively, depending on the variant of the recommender A. As we see in Fig. 3.3, the fuzzy sets have semantic meaning as: low, medium, high. In case of five fuzzy sets instead of three, the meaning can be e.g.: very low, low, medium, high, very high. The semantic meaning constitutes the main difference between both the WM-T and NIT-T that have the semantic interpretation, and the NIT-S without such an interpretation for the consequent constant values.

However, the core of this approach is the assumption that v_k, for $k = 1, 2, \ldots, N$ can be optimized without loss of interpretation. Therefore, they are optimized within certain intervals, as depicted at the bottom of Fig. 3.2.

The intervals are determined individually for every v_k, for $k = 1, 2, \ldots, N$, where N is the number of fuzzy IF-THEN rules; hence $v_k \in [v_{k,min}; v_{k,max}]$. Values $v_{k,min}$ and $v_{k,max}$ can be obtained by use of the following formulas:

$$v_{k,min} = arg \left\{ \min_{j=1,2,\ldots,M} \{\mu_{B^k}(y^{(j)})\} \right\} \tag{3.2}$$

$$v_{k,max} = arg \left\{ \max_{j=1,2,\ldots,M} \{\mu_{B^k}(y^{(j)})\} \right\} \tag{3.3}$$

for $R^{(k)}$ with $\max\{\tau_k\}$, $k = 1, 2, \ldots, N$, where τ_k is given by (2.8), and $(\mathbf{x}^{(j)}; y^{(j)})$, for $j = 1, 2, \ldots, M$, denotes the input-output data pairs.

Thus, the intervals are different for every rules $R^{(k)}$, $k = 1, 2, \ldots, N$, and determined by Eqs. (3.2) and (3.3), for the rule with maximum rule activation level for $(\mathbf{x}^{(j)}; y^{(j)})$.

The optimization of v_k, for $k = 1, 2, \ldots, N$, within intervals $[v_{k,min}; v_{k,max}]$ can be realized by various methods, including evolutionary algorithms. In [25], the GWO [8, 16] is applied.

The GWO algorithm mimics the leadership hierarchy and hunting mechanism of grey wolves in nature. Four types of grey wolves, such as alpha, beta, delta, and omega, are employed for simulating the leadership hierarchy. In addition, three main steps of hunting, searching for prey, encircling prey, and attacking prey, are implemented to perform optimization.

For details concerning the GWO algorithm applied in order to obtain the WM-T+S, NIT-T+S, NIT-S+S, see [25].

3.1.3 Systems Performance Evaluation

All the systems, WM-T, NIT-T, NIT-S, and their optimized versions, WM-T+S, NIT-T+S, NIT-S+S, presented in Sect. 3.1.2, have been tested on the MovieLens 10M dataset. The experiments were conducted for $m = 3$ and $m = 5$ where m denotes the number of fuzzy sets (linguistic values) of B^k, $k = 1, 2, \ldots, N$. Gaussian membership functions of these fuzzy sets were employed; see Fig. 3.3. The GWO optimization method was used with the following parameters: population size = 16, number of iterations = 100. The data from the MovieLens database was prepared for 100 users that rated more than 30 movies.

In order to verify the results of the simulations, 10-fold cross-validation was applied. Different data pairs were used for learning and testing, and also for creating the fuzzy rule base (by the WM and NIT methods). The following measures asses the systems performance: $rmse$—that denotes the RMSE (Root Mean Square Error), $accuracy$—the predicted value was rounded to the user rate, and thus ten different classes were obtained, and yes/no—the output value was set to class 1 if the prediction was lower than the average rate and to class 2 otherwise (introduced in [24]).

The simulation results in details are also presented in [25]. The following conclusions can be derived from the experiments:

- The optimization of the consequent values **v** within the specified intervals allows increasing the systems' accuracy (for the WM-T+S, NIT-T+S, and NIT-S+S).
- The best $rmse$, $accuracy$, and yes/no were obtained for the NIT-S+S system where initial consequent values **v** have been determined by the NIT method.
- The proposed approach allows achieving very high yes/no recommendation accuracy for testing data and high classification accuracy of predicting the exact user rate of the movie.
- Better accuracy is for $m = 5$ than for $m = 3$. However, for $m > 7$ we do not see significant improvement. Moreover, more linguistic labels of the consequent fuzzy sets make the interpretability of the system worse.

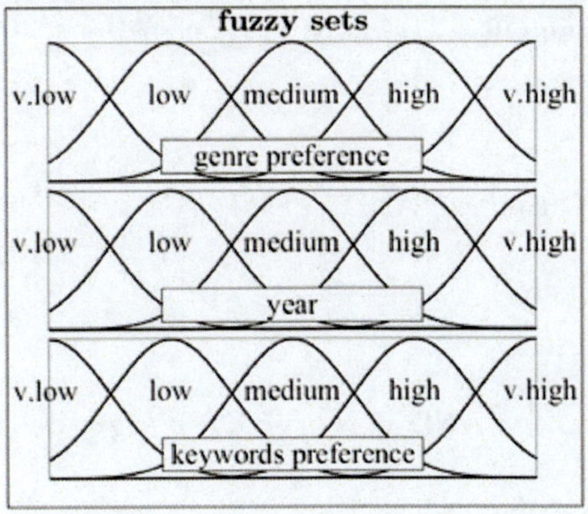

detailed results for selected user
$rmse$ = 0.17194 (lrn), 0.21830 (tst)
$accuracy$ = 87.9% (lrn), 84.3% (tst)
$yesno$ = 100.0% (lrn), 99.3% (tst)

fuzzy rules				
	IF			THEN
k	genre preference	year	keywords preference	user rate
	IS			=
1	v.low	medium	v.low	2.01
2	high	v.low	high	3.99
3	high	medium	medium	2.99
4	high	high	low	2.86
5	high	high	medium	3.02
6	high	high	high	4.15
7	high	v.high	low	2.85
8	high	v.high	medium	3.45
9	high	v.high	high	4.15
10	v.high	low	medium	3.07
11	v.high	medium	low	2.93
12	v.high	medium	medium	2.92
13	v.high	medium	high	5.00
14	v.high	high	medium	3.00
15	v.high	high	high	4.61
16	v.high	high	v.high	5.00
17	v.high	v.high	low	2.32
18	v.high	v.high	medium	3.03
19	v.high	v.high	high	4.65
20	v.high	v.high	v.high	5.00

Fig. 3.4 Examples of fuzzy rules of recommender A, for $m = 5$

- It is obvious that the number of rules generated for $m = 5$ is bigger than for $m = 3$. The interpretation of the recommender performance is easier for the smaller number of the rules.
- It is worth noticing that although the interpretation is more difficult for more rules, the explanation of specific recommendations is produced only based on the rules with higher levels of activation, not taking into account all of the rules.

The proposed approach allows achieving high accuracy with a reasonable number of interpretable fuzzy rules. The use of the ZO-TSK and the optimization of values v_k, for $k = 1, 2, \ldots, N$, significantly improves the results. The experiments showed that the GWO gives better accuracy without losing the interpretability of the system. The ZO-TSK system can be effectively applied as a content-based recommender that provides accurate results with interpretability, transparency, and explainability.

Figure 3.4 illustrates an example of fuzzy IF-THEN rules for the NIT-S+S system, for the user with id = 127 that rated 192 movies, for $m = 5$. It is easy to provide an explanation concerning recommendations based on these rules.

By use of the rules portrayed in Fig. 3.4, we can formulate the following IF-THEN rules with fuzzy numbers (see e.g. [6]) in the consequent parts:

$$\textbf{IF } x_1 \text{ is } low \text{ AND } x_2 \text{ is } medium \text{ AND } x_3 \text{ is } low \textbf{ THEN } y \approx 2$$

$$\textbf{IF } x_1 \text{ is } high \text{ OR } v.high \text{ AND } x_2 \text{ is } high \text{ OR } v.high \text{ AND } x_3 \text{ is } medium$$

$$\textbf{THEN } y \approx 3$$

$$\textbf{IF } x_1 \text{ is } high \text{ AND } x_2 \text{ is } high \text{ OR } v.high \text{ AND } x_3 \text{ is } high$$

$$\textbf{THEN } y \approx 4$$

$$\textbf{IF } x_1 \text{ is } v.high \text{ AND } x_2 \text{ is } high \text{ OR } v.high \text{ AND } x_3 \text{ is } high \text{ OR } v.high$$

$$\textbf{THEN } y \approx 5$$

where ≈ 2, ≈ 3, ≈ 4, ≈ 5 denote fuzzy numbers with the meaning described as "approximately" 2, 3, 4, 5, respectively.

It can be concluded that the exemplary neuro-fuzzy recommenders presented in this section are interpretable, allow providing an explanation based on the rules, and also ensure high accuracy of the performance.

3.2 Recommender B

This section presents the main algorithm proposed for design explainable recommender systems, denoted as WM, WO-C1, WR-C1, WO-C2, and WR-C2. As the recommender B, the Mamdani system with the WM rule generation method is employed. Hence, it differs from the recommender A because it is not considered as the ZO-TSK system. Moreover, the rule weights have been introduced. Computer simulations that illustrate the performance of all the variants of the recommender B have been realized based on the MovieLens 10M dataset, and the results are discussed in Sect. 3.2.4.

3.2.1 Introduction to the Proposed Recommender B

There are various methods applied in order to improve the performance of neuro-fuzzy systems, from introducing weights of the fuzzy rules (see e.g. [12]), through reducing the number of the fuzzy rules (see e.g. [4]), to optimizing a fuzzy system, usually by modifying fuzzy set parameters (see e.g. [14]), or rule consolidation (see e.g. [20]), and using the collaborative fuzzy clustering (see e.g. [19]).

In this section, the recommeder B based on the neuro-fuzzy approach is proposed that has the following characteristics:

- The WM method of rule generation allows producing interpretable and simple IF-THEN rules.
- The optimization of the rule weights does not modify parameters of the fuzzy sets in order to keep the interpretability of the rules.
- The rule reduction is applied not only to make the system simpler but also in order to decrease the error that expresses the system's accuracy.
- It allows checking how the optimization of the weights works in case of different levels of the rule reduction.
- It employs the standard ES (evolutionary strategy), see [28], as a method of weight optimization.
- It includes a version where the weights are rounded in order to increase the interpretability of the system.
- It uses isocriterial lines in order to evaluate an optimal balance between the system error (accuracy) and interpretability.

The WM, WO-C1, WR-C1, WO-C2, and WR-C2 are variants of the recommender B that refer to the differences concerning optimization of the weights, rounding their values, etc.; the details are explained in Sect. 3.2.2.

The feature encoding of the nominal values of the data is realized in the same way as described in Sect. 3.1.1; see Fig. 3.1 as the example.

3.2.2 *Description of the Recommender B*

In the approach presented in this section, system optimization is employed after a successive reduction of fuzzy rules. In this case, optimization refers to the optimized values of the weights. The rule reduction procedure is applied in order to simplify the rule base and increase the accuracy of the system.

When the weights of the rules are introduced, the mathematical formula that describes the neuro-fuzzy system, instead of (2.9), takes the following form:

$$\overline{y} = \frac{\sum_{k=1}^{N} w_k \ \tau_k \ \overline{y}^k}{\sum_{k=1}^{N} w_k \ \tau_k} \tag{3.4}$$

Although Eq. (3.4) has been applied in the literature in various problems of classification and modeling (see e.g. [1, 12, 13, 17, 22]), for the first time it was used in [26] in the context of designing fuzzy explainable recommenders.

Despite the fact that the usefulness of rule weighs is discussed in the literature (see e.g. [17, 29]), we find this issue interesting from interpretability and explainability point of view with regard to recommender systems. In general, the weights can be interpreted as the rule importance in the sense of expressing the number of data items in a dataset described by this rule. The more data items match the rule, the more important it is.

Table 3.1 Variants of the recommender B

Variants	Cases	Weight values
WM	–	Not optimized
WO-C1	C1	Optimized and reset
WR-C1	C1	Optimized, rounded, and reset
WO-C2	C2	Optimized and do not reset
WR-C2	C2	Optimized, rounded, do not reset

This approach allows significantly reducing the number of IF-THEN rules. In order to reduce the rules, the method of successively removing the least beneficial fuzzy rules was employed. To find the least beneficial rule, particular rules are turned off and on, one by one from the system, and after each change, the error (RMSE) is determined. As the least beneficial rule, this one is considered that causes the lowest RMSE error after its removal. For the details, see [26], and Algorithm 1.

During the optimization of the rule weights, it is assumed that only the weights (initialized equally by default) can be modified. For this purpose, the standard evolutionary strategy, known as ES $(\mu + \lambda)$, is applied; see e.g. [23]. However, any optimization algorithm can be used instead. This approach allows keeping a semantically interpretable form of fuzzy sets, and in addition, increases the system accuracy.

Moreover, this approach is more legible than the interpretation of systems in which fuzzy sets become uninterpretable after the optimization of their parameters (see e.g. [14]). The optimized system is denoted as WO.

Two approaches are considered: C1—where values of the weights are reset each time when the fuzzy rule is removed, and C2—where values of the weights are remembered after the optimization and do not reset.

The system with values of the weights that are rounded (to one decimal place) after the optimization is also considered. The variant with the rounded weight values is denoted as WR, while WO means the system with optimized weights (not rounded values). Table 3.1 explains notations of the recommendation systems as WM, WO-C1, WR-C1, WO-C2, and WR-C2.

The optimization and evaluation of the recommender are repeated after the removal of each fuzzy rule, with the goal of finding the best reduction level for which the weight values give the best system performance. This approach is realized, according to Algorithm 1.

It is worth adding that, as part of the tests, the results for a given number of rules and a specific reduction level were saved and then averaged. This is due to the fact that in the recommendation systems, the datasets for each user are different, so a different number of fuzzy rules is created by the use of the WM method. The recommender performance is evaluated in Sect. 3.2.4.

Algorithm 1 Design and reduction of explainable recommender systems

```
for all users do
  create N fuzzy rules using the WM method
  set system weights to equal values
  while N > 3 do
    evaluate system (WM)
    optimize system weights using ES
    remember system weights (only for variant C2)
    evaluate system (WO)
    round system weights
    evaluate system (WR)
    reset system weights (only for variant C1)
    for k = 1 to N do
      temporarily remove k-th fuzzy rule
      evaluate the system and remember the error
      include the removed fuzzy rule
    end for
    remove the least beneficial fuzzy rule
  end while
end for
```

3.2.3 Criteria of Balance Evaluation Between Recommender Accuracy and Interpretability

The performance of recommender B is enhanced by three criteria that allow evaluating an optimal balance in terms of accuracy and interpretability. It is worth emphasizing that the method illustrated as the pseudocode of Algorithm 1 generates different system accuracy for a different level of the rule reduction. Hence, choosing the optimal balance of the accuracy-interpretability is not a trivial task. Therefore, the isocriterial lines and criteria for model evaluation with regard to their complexity are beneficial.

The following criteria (see e.g. [3, 30]) have been employed to evaluate the recommender B: the Akaike Information Criterion (AIC), the Final Prediction Error (FPE), and the Schwartz criterion.

The AIC is defined in the following way:

$$AIC = M \ln Q + 2p \tag{3.5}$$

where M stands for the number of dataset samples (in this paper this value was set to the average number of the dataset samples generated for all users), Q denotes the system error, p means parameters that are optimized in the system (it is equal to the number of weights and analogously the number of fuzzy rules).

The FPE is expressed as follows:

$$FPE = Q \frac{M n + p}{M n - p} \tag{3.6}$$

where n denotes the number of system inputs.

The Schwarz criterion is defined by the following equation:

$$S = M \ln Q + p \, lnM \tag{3.7}$$

The isocriterial lines represent fixed values of these criteria, with different values of the system errors and the number of system parameters. This approach allows solving the problem of compromise between the system error and the number of optimized parameters describing the system. The points located on the isocriterial lines, with the smallest criterion values, characterize systems that are called sub-optimal. The sub-optimal systems provide the smallest values of statistical criteria within the examined structures (in this case differing in the number of system rules).

3.2.4 Recommenders Performance Evaluation

In the dataset, the rates of objects (movies) take numerical values. The rules obtained by the WM method for the Mamdani system are employed to solve regression tasks (see e.g. [4]), with the goal of the user rate prediction. This allows determining the RMSE error and also obtaining a more accurate classification of recommended objects.

In addition to the RMSE, accuracy (ACC) of predictions of the exact user's rate was calculated. In order to do this, the system's output values are rounded to the possible values of the object's rate, and in this way, the standard classification error (ACC) could be obtained (see e.g. [15]). There is also a classification error (YES/NO) for checking whether an item (object) should be recommended (if the value of its recommendation is more than half of possible values) or not; see the yes/no in Sect. 3.1.3.

Simulations that illustrate the performance of the recommender B are described in this section. The results of the simulations verify the effectiveness of the recommendation system for the MovieLens 10M dataset.

In the simulations, the cases presented in Table 3.1 were tested. In C1, weight values are reset after each reduction of a fuzzy rule, while in C2, weight values are remembered after optimization and do not reset (see Algorithm 1).

The number of fuzzy sets (linguistic values), L_i, for each attribute corresponding to linguistic variables x_i, for $i = 1, 2, \ldots, n$, has been chosen as the Gaussian-type fuzzy sets, for $L_i = 5$. The following parameters of the ES algorithm were set: population size = 32, number of iterations = 100, evaluation function = RMSE.

Three inputs, $n = 3$, have been considered, with regard to the movie recommendations: genre preference), year (numeric values), and also keywords preference. Moreover, datasets were prepared for the first 100 users that rated more than 30 movies from the database. As mentioned earlier, the nominal values have been encoded according to the method described in Sect. 3.1.1. The datasets prepared for the first 100 users that rated more than 30 movies from the database have been used.

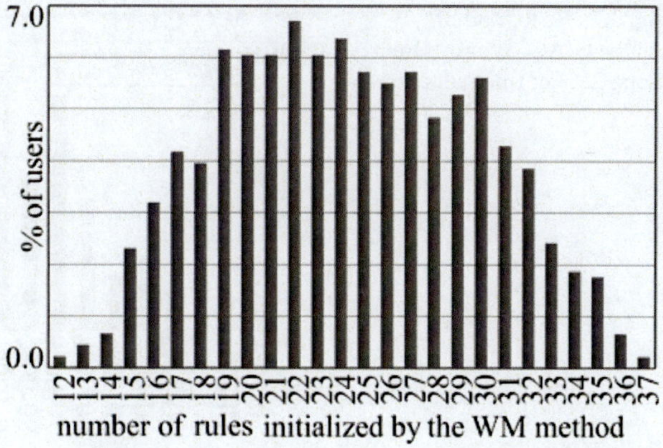

Fig. 3.5 Histogram of the initial number of rules generated by use of the WM method

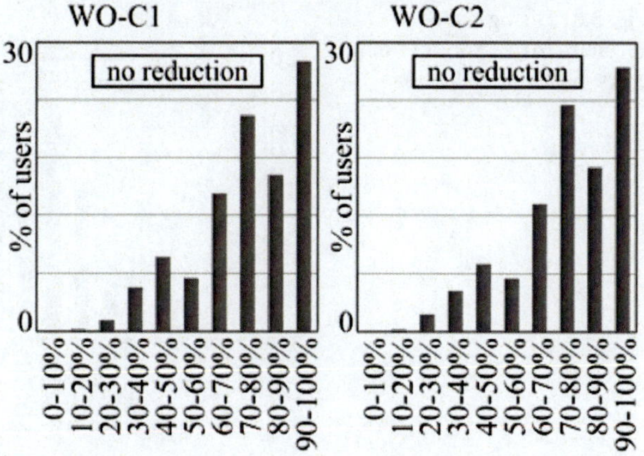

Fig. 3.6 Histogram of accuracy (ACC) for different users; no rule reduction

As a testing method, 10-fold cross-validation was applied, and only learning data pairs had been used for creating fuzzy rules by the WM method. The remaining data played the role of the testing data pairs.

In the recommendation systems, rating predictions for new data (object attribute values) are important. Therefore, the error values for the testing data are crucial in the result comparison. Hence, most of the conclusions are focused on this analysis.

Figure 3.5 portrays the histogram that illustrates the initial number of IF-THEN rules generated by the WM method. Depending on a user and the number of rated objects, the WM algorithm produces different numbers of the fuzzy rules, with an average of 24 rules for the proposed features' encoding. The number of rules for particular users differs from 12 to 37.

Histograms of accuracy (ACC) for different users are presented in Figs. 3.6, 3.7, 3.8 and 3.9, for the WO-C1 and WO-C2 systems.

Detailed simulation results are included in Tables 3.2, 3.3 and 3.4, for the RMSE, ACC, and YES/NO, respectively.

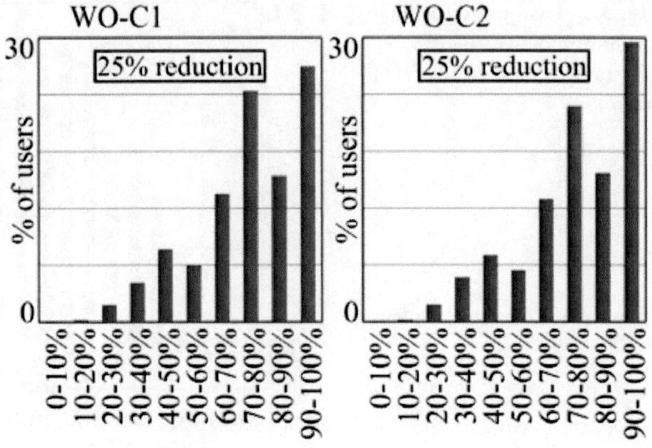

Fig. 3.7 Histogram of accuracy (ACC) for different users; 25% of rule reduction

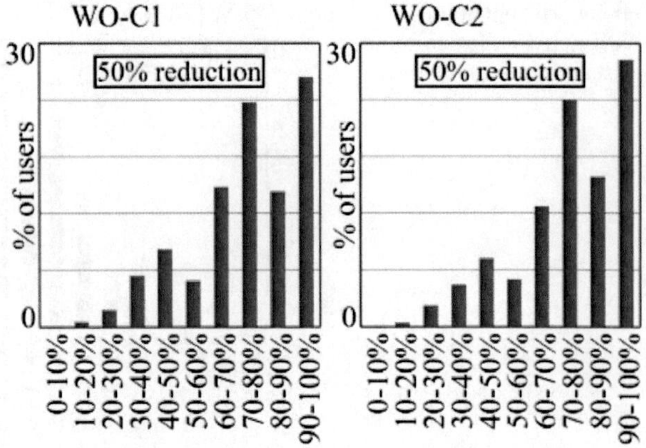

Fig. 3.8 Histogram of accuracy (ACC) for different users; 50% rule reduction

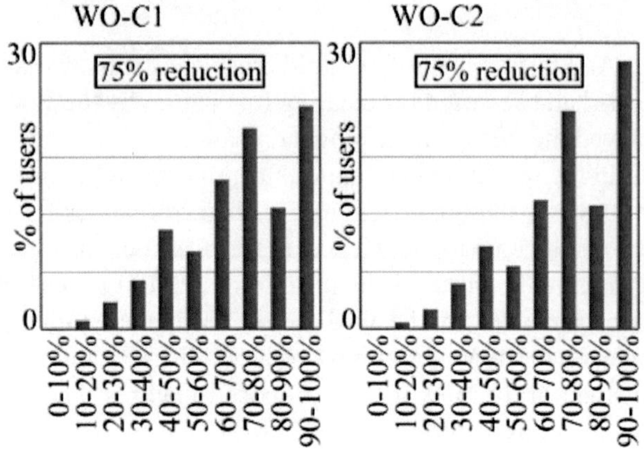

Fig. 3.9 Histogram of accuracy (ACC) for different users; 75% of rule reduction

Table 3.2 Average RMSE for all users in terms of % of reduced rules (RR)

% of	Learning data					Testing data				
RR	WM	WO-C1	WR-C1	WO-C2	WR-C2	WM	WO-C1	WR-C1	WO-C2	WR-C2
0%	0.2199	0.1425	**0.1485**	0.1423	**0.1472**	0.3096	0.2670	0.2688	0.2677	0.2701
4%	0.2302	0.1605	0.1637	0.1513	0.1580	0.3023	**0.2641**	**0.2669**	0.2619	**0.2648**
8%	0.2106	0.1518	0.1582	0.1408	0.1560	0.2980	0.2674	0.2698	0.2658	0.2719
12%	0.1915	0.1395	0.1470	0.1283	0.1465	0.2910	0.2670	0.2706	0.2635	0.2724
16%	0.1908	0.1448	0.1511	0.1316	0.1534	0.2890	0.2688	0.2713	0.2638	0.2734
20%	0.1876	0.1462	0.1535	0.1329	0.1595	0.2876	0.2693	0.2713	0.2662	0.2765
24%	0.1807	**0.1421**	0.1497	0.1282	0.1576	0.2858	0.2669	0.2701	**0.2600**	0.2731
28%	0.1801	0.1429	0.1527	0.1279	0.1638	0.2932	0.2755	0.2791	0.2682	0.2863
32%	0.1789	0.1438	0.1531	0.1290	0.1684	0.2914	0.2748	0.2777	0.2683	0.2906
36%	**0.1775**	0.1435	0.1540	0.1285	0.1716	**0.2843**	0.2698	0.2741	0.2637	0.2848
40%	0.1801	0.1463	0.1580	0.1299	0.1804	0.2896	0.2750	0.2795	0.2692	0.2937
44%	0.1803	0.1465	0.1583	0.1299	0.1794	0.2949	0.2779	0.2823	0.2713	0.2956
48%	0.1797	0.1455	0.1594	**0.1278**	0.1869	0.2917	0.2739	0.2782	0.2683	0.2958
52%	0.1892	0.1519	0.1663	0.1325	0.1910	0.2971	0.2776	0.2826	0.2721	0.2994
56%	0.1956	0.1561	0.1741	0.1353	0.2002	0.3026	0.2793	0.2858	0.2738	0.3039
60%	0.1993	0.1592	0.1830	0.1375	0.2120	0.3048	0.2845	0.2935	0.2747	0.3057
64%	0.2143	0.1708	0.1955	0.1467	0.2240	0.3209	0.2941	0.3042	0.2802	0.3110
68%	0.2292	0.1826	0.2139	0.1570	0.2429	0.3272	0.2968	0.3067	0.2806	0.3189
72%	0.2442	0.1963	0.2321	0.1691	0.2582	0.3291	0.2980	0.3127	0.2813	0.3280
76%	0.2691	0.2179	0.2585	0.1886	0.2902	0.3532	0.3192	0.3364	0.2932	0.3448
80%	0.3025	0.2453	0.2967	0.2168	0.3166	0.3715	0.3305	0.3532	0.2978	0.3538
84%	0.3450	0.2826	0.3513	0.2465	0.3504	0.3892	0.3415	0.3758	0.3024	0.3638
88%	0.3945	0.3289	0.3950	0.2815	0.3776	0.4229	0.3680	0.4089	0.3233	0.3830
92%	0.4406	0.3642	0.4229	0.3045	0.4021	0.4685	0.3943	0.4311	0.3344	0.4035

Analyzing Tables 3.2 and 3.3, we see that 32–36% of the rules, generated by the WM method, can be removed resulting in better performance of the system on both learning and testing data; with regard to the average RMSE and ACC. However, in Table 3.4, where the YES/NO classification error is considered, the same result has been obtained only for the learning data. It is worth noticing that the value of the YES/NO error in the case of 36% rule reduction does not differ much from the value of this error for the case of 0%, i.e., no rule reduction. This means that 36% of the rules can be removed but not more in all the cases of the WM system.

The isocriterial lines for the AIC, FPE, and Schwarz criteria, for the learning and testing data, in comparison of the WO-C1 and WO-C2 systems, are illustrated in Figs. 3.10 and 3.11. Values of these criteria, as well as the RMSE, are presented in Tables 3.5 and 3.6, for different number of rules (and also weights), p. From these results, we can know the number of rules that can be removed with an optimal balance in terms of system accuracy and interpretability.

Table 3.3 Average ACC for all users in terms of % of reduced rules (RR)

% of	Learning data					Testing data				
RR	WM	WO-C1	WR-C1	WO-C2	WR-C2	WM	WO-C1	WR-C1	WO-C2	WR-C2
0%	77.699	91.091	90.600	91.019	90.607	67.149	**76.721**	**76.469**	76.463	75.903
4%	76.523	89.049	88.595	90.213	89.533	67.095	75.701	75.783	76.087	75.969
8%	79.674	89.838	89.337	91.279	90.114	68.622	75.684	75.424	76.433	75.760
12%	82.977	**91.313**	**90.892**	92.679	**91.370**	70.778	76.154	75.830	76.810	75.908
16%	83.371	90.870	90.444	92.481	90.734	71.440	75.601	75.407	76.254	75.303
20%	83.562	90.565	90.018	92.221	90.126	71.341	75.424	75.047	76.408	75.142
24%	84.641	90.913	90.450	92.629	90.446	71.924	76.007	75.488	**77.361**	**76.089**
28%	85.113	91.018	90.458	**92.871**	90.215	72.212	75.628	75.451	76.992	75.490
32%	**85.344**	90.832	90.217	92.753	89.784	72.393	75.514	75.222	76.691	74.724
36%	85.340	90.644	89.970	92.654	89.499	**72.715**	75.504	75.011	76.682	74.880
40%	84.921	90.408	89.694	92.568	88.833	72.458	75.258	74.652	76.504	74.453
44%	84.882	90.415	89.615	92.602	88.959	71.478	74.953	74.275	76.147	73.934
48%	84.744	90.372	89.512	92.774	88.731	72.012	75.177	74.604	76.137	73.853
52%	83.361	89.594	88.683	92.266	88.184	71.875	75.049	74.671	76.568	74.182
56%	82.466	89.009	87.846	91.903	87.524	70.927	75.035	74.588	76.575	74.082
60%	81.833	88.428	87.055	91.445	86.700	70.369	74.356	73.855	76.099	73.787
64%	80.098	87.106	85.554	90.440	85.515	68.854	73.377	72.607	75.521	72.940
68%	78.280	86.052	84.154	89.407	83.852	68.518	73.879	73.221	75.819	73.069
72%	76.838	84.565	82.632	88.170	82.539	68.159	73.075	72.258	75.639	72.642
76%	74.454	82.450	80.187	86.261	80.290	65.625	71.480	70.541	74.488	71.682
80%	71.005	79.897	77.270	83.562	78.153	64.340	70.986	69.363	74.530	71.608
84%	67.335	75.982	72.789	80.602	74.940	62.052	68.947	67.010	73.520	70.176
88%	62.994	71.200	68.071	77.023	71.904	58.543	66.360	64.570	71.274	67.985
92%	59.315	68.213	65.502	74.907	70.161	54.622	64.517	62.078	70.704	66.941

Analyzing results included in the tables, we observe that the systems with the weights perform better than the WM recommender.

Comparing the different variants of the system with the weights, we can conclude that generally, the WO-C2 gives better results. This is the system with optimized weights (and not rounded) in the case when the weights have not been reset. This means that the procedure of reducing the rules concerns always the same system (not changed by resetting values of the weights).

It seems obvious that better performance is obtained for the system with optimized weights and not rounded values. However, this is also a compromise between accuracy and interpretability.

Table 3.4 Average YES/NO for all users in terms of % of reduced rules (RR)

% of	Learning data					Testing data				
RR	WM	WO-C1	WR-C1	WO-C2	WR-C2	WM	WO-C1	WR-C1	WO-C2	WR-C2
0%	98.852	99.387	99.286	99.397	**99.316**	97.659	97.981	97.961	98.049	97.948
4%	98.988	99.358	99.309	99.370	99.301	**98.242**	**98.541**	**98.552**	**98.705**	**98.649**
8%	99.085	99.368	99.288	99.422	99.224	98.146	98.291	98.293	98.382	98.184
12%	99.155	99.399	**99.329**	99.431	99.193	97.971	97.807	97.789	98.046	97.814
16%	99.121	99.371	99.284	99.399	99.043	98.036	97.986	97.968	98.222	98.005
20%	99.115	99.346	99.217	99.403	98.924	97.962	98.063	98.079	98.170	97.945
24%	99.152	99.350	99.252	99.412	98.998	97.806	97.960	97.940	98.149	97.768
28%	99.147	99.344	99.210	99.414	98.763	97.772	97.884	97.838	98.250	97.794
32%	99.166	99.349	99.224	99.437	98.817	97.792	97.844	97.750	98.310	97.828
36%	**99.256**	**99.419**	99.278	**99.489**	98.809	97.808	97.847	97.769	98.189	97.749
40%	99.188	99.355	99.205	99.430	98.398	97.554	97.811	97.788	98.120	97.424
44%	99.106	99.301	99.167	99.392	98.727	97.599	97.940	97.935	98.224	97.635
48%	99.164	99.345	99.184	99.419	98.511	97.681	97.910	97.839	98.143	97.259
52%	99.133	99.341	99.165	99.428	98.640	97.655	98.036	97.996	98.430	97.504
56%	99.129	99.328	99.088	99.410	98.529	97.647	98.009	97.905	98.364	97.471
60%	99.062	99.264	99.020	99.356	98.163	97.453	97.846	97.595	97.909	97.238
64%	98.907	99.171	98.887	99.302	98.379	97.139	97.525	97.379	98.037	97.544
68%	98.792	99.120	98.624	99.337	98.022	97.152	97.654	97.472	97.948	97.065
72%	98.618	98.980	98.377	99.262	97.851	97.291	97.610	97.165	98.012	96.877
76%	98.150	98.694	98.052	99.091	97.331	96.439	97.159	96.767	97.828	96.400
80%	97.663	98.352	97.428	98.842	96.898	96.199	97.057	96.393	97.573	95.790
84%	96.953	97.760	96.104	98.576	96.211	95.478	96.538	95.407	97.324	95.066
88%	96.563	97.287	95.385	98.422	95.765	95.564	96.459	94.891	97.697	94.976
92%	96.306	97.175	95.482	98.182	95.356	95.588	96.577	95.155	97.478	94.695

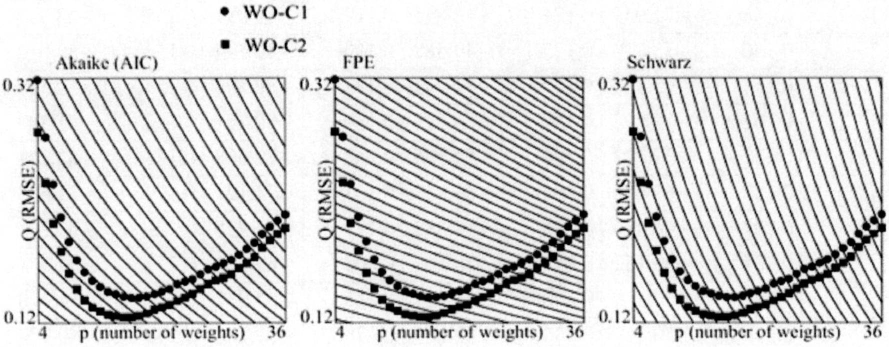

Fig. 3.10 Isocriterial lines for learning data; circles—WO-C1, squares—WO-C2

Table 3.5 Evaluation of the WO recommenders, for learning data

p	WO-C1				WO-C2			
	RMSE	AIC	FPE	Schwarz	RMSE	AIC	FPE	Schwarz
36	0.209	−7.927	0.337	61.619	0.198	−10.650	0.319	58.896
35	0.204	−11.108	0.325	56.506	0.193	−14.007	0.307	53.607
34	0.199	−14.378	0.312	51.304	0.188	−17.250	0.295	48.432
33	0.193	−17.810	0.300	45.940	0.182	−21.027	0.281	42.723
32	0.186	−21.873	0.284	39.946	0.174	−25.159	0.266	36.659
31	0.181	−25.184	0.273	34.702	0.169	−28.662	0.255	31.225
30	0.176	−28.475	0.263	29.480	0.165	−31.851	0.246	26.104
29	0.172	−31.919	0.252	24.104	0.160	−35.370	0.235	20.653
28	0.169	−34.766	0.244	19.325	0.157	−38.293	0.228	15.798
27	0.166	−37.447	0.238	14.712	0.155	−41.215	0.221	10.944
26	0.164	−40.245	0.231	9.982	0.152	−44.120	0.214	6.108
25	0.160	−43.477	0.222	4.818	0.148	−47.511	0.206	0.784
24	0.156	−46.688	0.214	−0.324	0.144	−51.000	0.197	−4.636
23	0.154	−49.434	0.208	−5.002	0.141	−53.811	0.191	−9.379
22	0.152	−52.208	0.203	−9.707	0.138	−56.835	0.185	−14.335
21	0.148	−55.310	0.196	−14.742	0.135	−59.978	0.178	−19.410
20	0.146	−58.138	0.190	−19.501	0.133	−62.931	0.173	−24.295
19	0.144	−60.752	0.185	−24.047	0.131	−65.756	0.168	−29.051
18	0.142	−63.454	0.180	−28.681	0.128	−68.664	0.163	−33.891
17	0.141	−65.943	0.176	−33.102	0.126	−71.619	0.158	−38.778
16	**0.141**	−67.811	0.174	−36.902	0.125	−73.909	0.155	−43.000
15	0.142	−69.412	0.173	−40.435	**0.125**	−75.875	0.153	−46.898
14	0.144	−70.893	**0.173**	−43.848	0.126	−77.662	**0.151**	−50.617
13	0.146	−71.969	0.174	−46.855	0.127	−79.091	0.151	−53.977
12	0.150	−72.770	0.175	−49.588	0.130	−80.194	0.152	−57.012
11	0.155	**−73.184**	0.179	−51.934	0.133	**−80.806**	0.154	−59.556
10	0.162	−72.892	0.184	−53.574	0.139	−80.701	0.158	−61.383
9	0.172	−71.723	0.194	**−54.336**	0.148	−79.608	0.166	**−62.221**
8	0.187	−69.512	0.208	−54.057	0.161	−77.127	0.179	−61.673
7	0.207	−66.410	0.226	−52.887	0.179	−73.725	0.196	−60.203
6	0.234	−62.152	0.253	−50.561	0.202	−69.562	0.219	−57.971
5	0.272	−56.470	0.290	−46.811	0.235	−63.937	0.250	−54.278
4	0.319	−50.329	0.336	−42.601	0.276	−57.706	0.291	−49.979

Table 3.6 Evaluation of the WO recommenders, for testing data

p	WO-C1				WO-C2			
	RMSE	AIC	FPE	Schwarz	RMSE	AIC	FPE	Schwarz
36	0.261	3.529	0.422	73.075	0.258	2.977	0.417	72.523
35	0.248	−1.097	0.395	66.517	0.248	−1.205	0.394	66.409
34	**0.244**	−3.937	0.383	61.745	**0.245**	−3.678	0.385	62.004
33	0.252	−4.352	0.390	59.398	0.250	−4.720	0.387	59.031
32	0.261	−4.448	0.399	57.370	0.259	−4.969	0.395	56.850
31	0.260	−6.671	0.392	53.215	0.261	−6.593	0.393	53.293
30	0.264	−7.955	0.393	50.000	0.261	−8.564	0.388	49.391
29	0.261	−10.490	0.383	45.533	0.257	−11.338	0.377	44.685
28	0.260	−12.640	0.377	41.451	0.258	−13.185	0.373	40.906
27	0.264	−13.922	0.377	38.237	0.260	−14.710	0.371	37.449
26	0.264	−15.828	0.373	34.399	0.260	−16.672	0.367	33.556
25	0.265	−17.701	0.369	30.595	0.263	−18.119	0.366	30.176
24	0.268	−19.075	0.368	27.289	0.264	−19.891	0.362	26.473
23	0.273	−20.241	0.369	24.191	0.269	−20.931	0.364	23.501
22	0.271	−22.547	0.362	19.953	0.266	−23.616	0.355	18.884
21	0.269	−25.057	0.354	15.511	0.264	−25.933	0.348	14.636
20	0.270	−26.846	0.351	11.791	0.264	−27.898	0.344	10.739
19	0.271	−28.534	0.348	8.171	0.265	−29.773	0.340	6.932
18	0.274	−29.968	0.347	4.804	0.269	−30.939	0.341	3.834
17	0.274	−32.094	0.342	0.747	0.268	−33.112	0.335	−0.271
16	0.277	−33.530	0.341	−2.621	0.269	−35.015	0.332	−4.106
15	0.278	−35.233	0.339	−6.255	0.270	−36.817	0.328	−7.840
14	0.281	−36.777	0.337	−9.731	0.272	−38.441	0.327	−11.396
13	0.282	−38.590	0.334	−13.476	0.274	−40.013	0.325	−14.899
12	0.281	−40.805	**0.328**	−17.623	0.275	−41.839	0.322	−18.657
11	0.286	−41.903	0.330	−20.653	0.278	−43.284	0.321	−22.034
10	0.289	−43.327	0.329	−24.009	0.279	−45.068	0.318	−25.750
9	0.292	−44.838	0.328	−27.452	0.277	−47.388	0.312	−30.001
8	0.300	**−45.332**	0.334	−29.878	0.281	−48.673	**0.312**	−33.218
7	0.315	−44.962	0.345	−31.440	0.288	−49.446	0.316	−35.923
6	0.329	−44.724	0.356	−33.133	0.299	**−49.577**	0.323	−37.986
5	0.351	−43.439	0.374	−33.780	0.311	−49.639	0.332	−39.980
4	0.373	−42.362	0.393	**−34.635**	0.327	−49.043	0.344	**−41.316**

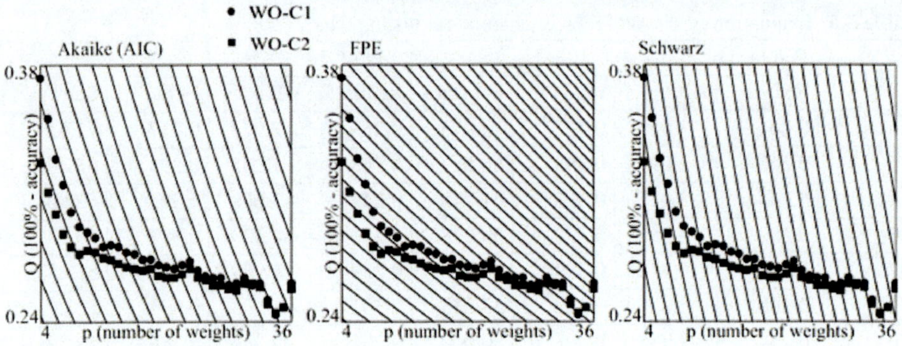

Fig. 3.11 Isocriterial lines for testing data; circles—WO-C1, squares—WO-C2

3.2.5 Interpretability and Explainability of the Recommender

The fuzzy rules, applied in the systems considered as recommender B, are semantically interpretable. Figures 3.12 and 3.13 portray examples of the IF-THEN rules for two users and three variants of the recommender. Fuzzy sets of rules are also illustrated. The rule base, after the optimal reduction of rules, is considered. With the semantic meaning of the fuzzy sets, the rules can easily explain the recommendations.

For example, the rule number 5 from Fig. 3.12—"*IF genre pref. is high and year is medium and keywords pref. is high THEN user rate is high*" can be used as an explanation. The recommendation is the conclusion of the rule, directly inferred from the antecedents and the input data. it is also possible to provide for a user the visualization of the particular fuzzy sets describing the meaning of the following terms: "*high genre pref.*", "*medium year*" and "*high keywords pref.*".

The number of rules is low and sufficient for the good accuracy of the recommender. The rule base allows explaining the performance of the system with regard to the recommendations. It is easy to present such an explanation for every user; in the same way as for two users based on Figs. 3.12 and 3.13.

3.3 Recommender C

In this section, four neuro-fuzzy recommenders, marked as WM, WM+W, WM+D, and WM+W+D, equipped with a novel feature encoding method, are studied, and their performance is evaluated in details, taking into account a compromise between systems' error and the number of rules.

fuzzy sets

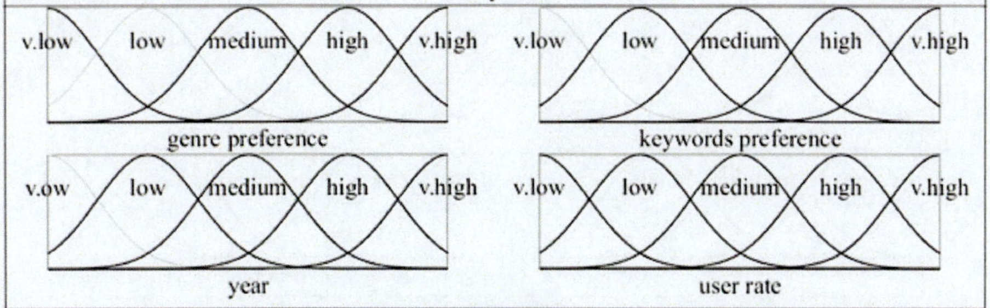

fuzzy rules

k	IF			THEN	rule weight		
	genre pref. is	year is	keywords pref. is	user rate is	WM	WO-C2	WR-C2
1	high	v.high	medium	medium	0.500	0.543	0.500
2	high	high	v.high	v.high	0.500	0.076	0.100
3	v.high	v.high	v.high	v.high	0.500	0.917	0.900
4	high	medium	v.high	v.high	0.500	0.604	0.600
5	high	medium	high	high	0.500	0.228	0.200
6	high	low	v.high	v.high	0.500	0.620	0.600
7	medium	v.high	low	v.low	0.500	0.591	0.600
8	v.high	v.high	high	high	0.500	0.465	0.500
9	high	v.high	low	v.low	0.500	0.580	0.600
10	high	low	medium	low	0.500	0.443	0.400
11	high	medium	medium	medium	0.500	0.296	0.300
12	high	high	low	v.low	0.500	0.149	0.100
13	high	low	high	high	0.500	0.516	0.500
14	medium	v.high	v.high	v.high	0.500	1.000	1.000
15	v.low	v.high	low	low	0.500	0.642	0.600
16	medium	v.high	medium	medium	0.500	0.345	0.300
system errors			RMSE	learning	0.247	0.210	0.212
				testing	0.344	0.330	0.328
			ACC	learning	75.1	82.0	80.1
				testing	57.1	65.8	61.9
			YES/NO	learning	99.4	99.4	99.4
				testing	100.0	100.0	100.0

Fig. 3.12 Examples of fuzzy rules for the user with id = 7

3.3.1 Nominal Attribute Values Encoding

Referring to Sect. 3.1.1, let us consider a database, $S = \{\mathbf{o}_1, \ldots, \mathbf{o}_M\}$, of M objects, $\mathbf{o}_j$, $j = 1, \ldots, M$, characterized by n attributes $A_{1,j}, \ldots, A_{n,j}$, and d_j that is the decision attribute. Hence, every object is expressed as follows: $\mathbf{o}_j = (A_{1,j}, \ldots, A_{n,j}, d_j)$.

Let us assume that i-th attribute of object $\mathbf{o}_j$, for $i = 1, \ldots, n$ and $j = 1, \ldots, M$, has nominal values. In the first step of the proposed method, we apply K_i-dimensional one-hot vector $X_{i,j} = [x_{i,j,1}, \ldots, x_{i,j,K_i}]^T$ where

fuzzy sets

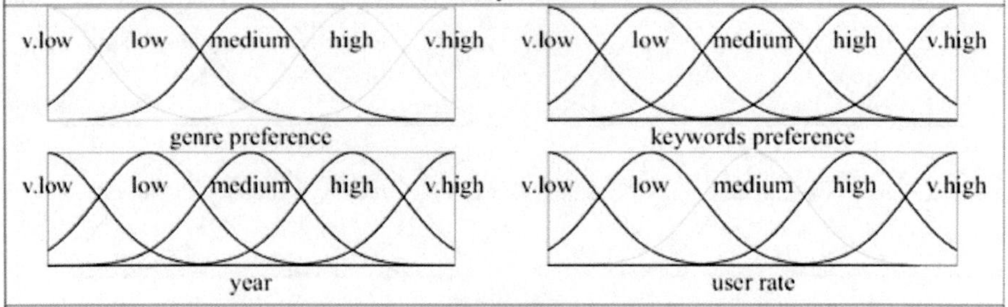

fuzzy rules

k	IF			THEN	rule weight		
	genre pref. is	year is	keywords pref. is	user rate is	WM	WO-C2	WR-C2
1	medium	medium	v.high	v.high	0.500	0.310	0.300
2	medium	v.high	medium	low	0.500	0.156	0.200
3	low	v.high	v.low	v.low	0.500	1.000	1.000
4	low	v.high	low	v.low	0.500	0.073	0.100
5	medium	v.high	v.high	v.high	0.500	1.000	1.000
6	medium	high	v.high	v.high	0.500	1.000	1.000
7	low	high	medium	low	0.500	0.108	0.100
8	low	v.low	v.high	v.high	0.500	0.450	0.500
9	medium	v.high	high	high	0.500	0.266	0.300
10	low	medium	high	high	0.500	0.050	0.100
11	low	v.high	medium	high	0.500	0.273	0.300
12	medium	low	low	low	0.500	0.056	0.100
system errors			RMSE	learning	0.171	0.113	0.141
				testing	0.188	0.173	0.159
			ACC	learning	86.7	88.9	88.9
				testing	100.0	100.0	100.0
			YES/NO	learning	97.7	100.0	100.0
				testing	100.0	100.0	100.0

Fig. 3.13 Examples of fuzzy rules for the user with id = 2

$$x_{i,j,h} = \begin{cases} 1 & \text{if } \ v_{i,h} \text{ is a value of attribute } A_{i,j}, \\ 0 & \text{otherwise}, \end{cases} \tag{3.8}$$

for $h = 1, \ldots, K_i$.

Let us consider the movie data, where i-th attribute of object $\mathbf{o}_j$ is *genre*, and it takes values from the set $\{comedy, drama, fiction, action\}$. Hence, for example, $X_{i,j} = [0, 1, 0, 0]^T$ if $A_{i,j} = \{drama\}$ but $X_{i,j} = [1, 0, 0, 1]^T$ if $A_{i,j} = \{comedy, action\}$.

In this section, it is proposed the way of transforming the nominal attribute values into numerical ones, simply by computing the Pearson's correlation coefficients between the appropriate one-hot vectors and the ratings (decision attribute), d_j, with the assumption of random variables, for $j = 1, \ldots, M$.

Thus, the Pearson's correlation coefficients, for $X_{i,j}$ and dj, that are the one-hot vector corresponding to i-th attribute and the rating, respectively, of randomly chosen object $\mathbf{o}_j$, characterized by the data item, are determined, for $j = 1, \ldots, M$, $i = 1, \ldots, n$, and $h = 1, \ldots, N_i$. The vector of correlation coefficients for i-th attribute, and h-th element of the one-hot vector, can be expressed as follows:

$$\rho_{i,j,h} = \frac{Cov\left(x_{i,j,h}, d_j\right)}{\sqrt{Var\left(x_{i,j,h}\right) Var\left(d_j\right)}}, . \tag{3.9}$$

In order to estimate correlation coefficient $\rho_{i,j,h}$, by use of the dataset, it is necessary to determine the following five estimators:

- average value of $x_{i,j,h}$, denoted as $\overline{x}_{i,h}$:

$$\overline{x}_{i,h} = \frac{1}{M} \sum_{j=1}^{M} x_{i,j,h}; \tag{3.10}$$

- average rating:

$$\overline{d} = \frac{1}{M} \sum_{j=1}^{M} d_j; \tag{3.11}$$

- variance of $x_{i,j,h}$:

$$\widehat{Var}\left(x_{i,j,h}\right) = \frac{1}{M-1} \sum_{j=1}^{M} \left(x_{i,j,h} - \overline{x}_{j,h}\right)^2; \tag{3.12}$$

- variance of rating:

$$\widehat{Var}(d) = \frac{1}{M-1} \sum_{j=1}^{M} \left(d_j - \overline{d}\right)^2; \tag{3.13}$$

- covariance between $x_{i,j,h}$ and rating:

$$\widehat{Cov}\left(x_{i,j,h}, d\right) = \frac{1}{M-1} \sum_{j=1}^{M} \left(x_{i,j,h} - \overline{x}_{j,h}\right)\left(d_j - \overline{d}\right), \tag{3.14}$$

where the unbiased estimators for variances and covariances are applied.

Finally, the correlation coefficients are estimated as follows:

$$\widehat{\rho}_{i,j,h} = \frac{\widehat{Cov}\left(x_{i,j,h}, d_j\right)}{\sqrt{\widehat{Var}\left(x_{i,j,h}\right) \widehat{Var}\left(d_j\right)}}, \tag{3.15}$$

for $i = 1, \ldots, n$, $j = 1, \ldots, M$, and $h = 1, \ldots, N_i$.

The estimators of correlation coefficients, given by (3.15), can be used in order to transform particular values of i-th attribute of object $\mathbf{o}_j$ to corresponding numerical values $a_{i,j}$, for $i = 1, \ldots, n$, and $j = 1, \ldots, M$.

Thus, the proposed procedure is composed of two steps. At first, the values of attribute $A_{i,j}$ are expressed as the one-hot vector, $X_{i,j} = \left[x_{i,j,1}, \ldots, x_{i,j,N_j}\right]^T$, by use of formula (3.8). Then, applying the estimators of correlation coefficient, given by (3.15), numerical values, $a_{i,j}$ are obtained from vector $X_{i,j}$, in the following way:

$$a_{i,j} = \frac{\sum_{h=1}^{N_i} x_{i,j,h} \, \overline{x}_{i,h} \, \widehat{\rho}_{i,j,h}}{\sum_{h=1}^{N_i} x_{i,j,h} \, \overline{x}_{i,h}}. \tag{3.16}$$

for $i = 1, \ldots, n$, $j = 1, \ldots, M$, and $h = 1, \ldots, N_i$, where $\overline{x}_{i,h}$ is gived by (3.10).

The numerical values $a_{i,j}$, determined according to formula (3.16), are applied in the neuro-fuzzy recommender systems.

3.3.2 *Various Neuro-Fuzzy Systems as the Proposed Recommender C*

The recommenders considered so far are based on the Mamdani (or TSK) fuzzy systems, with the rules generated by the use of the WM (or NIT) method; see Chap. 2. The neuro-fuzzy network presented in Fig. 2.1 constitutes the simplest form of different neuro-fuzzy architectures (see e.g. [21]). The basic triangular T-norm—product—that is commonly used, have been applied in recommenders A and B.

In this section, apart from the WM and WM+W, also the WM+D and WM+W+D recommenders are proposed. The WM denotes the Mamdani system with the fuzzy rules generated by the WM method. As a matter of fact, this is the same recommender, as considered in the previous sections (especially as recommender A). The WM+W is the system with the rule importance weights (see recommender B).

The new proposition concern the WM+D and WM+W+D that systems constructed by use of the Dombi T-norm, which is a parametric T-norm that in the simplest case is defined as follows:

$$T(x, y) = \begin{cases} 0 & \text{if } \; x = 0 \; \text{ or } \; y = 0, \\ \left(1 + \left(\left(\frac{1-x}{x}\right)^q + \left(\frac{1-y}{y}\right)^q\right)^{1/q}\right)^{-1} & \text{otherwise,} \end{cases} \tag{3.17}$$

where q is the Dombi T-norm parameter, and $q > 0$; see e.g. [23].

In the systems based on the Dombi T-norm, called WM+D and WMD+W+D, it is assumed that each rule has its own q_j parameter, for $j = 1, 2, \ldots, N$.

Fuzzy IF-THEN rules of the form (2.1) are applied in each variant of recommender C. However, instead of (2.8), the degree of rule activation (rule firing level), should now be expressed in the more general way, as follows:

Algorithm 2 Pseudocode of the proposed approach

```
for all users do
    create N fuzzy rules using WM method
    set system weights to equal values
    set Dombi parameters to 1
    while N > 4 do
        optimize system parameters using ES
        evaluate system
        for j = 1 to N do
            temporarily remove j-th fuzzy rule
            evaluate the system and remember the error
            include the removed fuzzy rule
        end for
        remove the least beneficial fuzzy rule
        calculate the Akaike criterion
    end while
end for
```

$$\tau_j = T\left\{\mu_{A_1^j}(\bar{x}_1), \mu_{A_2^j}(\bar{x}_2), \ldots, \mu_{A_n^j}(\bar{x}_n)\right\}, \tag{3.18}$$

for $j = 1, 2, \ldots, N$, where T denotes a T-norm that can be “product”, “min”, “Dombi” or other function as the T-norm (see e.g. [21, 23]).

The WM and WM+W systems refer to the case where the algebraic T-norm, $T(x, y) = xy$, most often used, is applied.

In the WM+W+D, the fuzzy rule weights and parameterized Dombi T-norm is employed. The idea of the proposed approach is included in Algorithm 2.

Of course, a proper selection of the rule weights and Dombi parameters improves the performance of recommenders. In order to optimize the system parameters, the evolutionary strategies (ES), as an optimization method, is employed in Algorithm 2.

Also, a simple mechanism for further reduction of the fuzzy rules, by rejection those that increase the system error, is included in Algorithm 2. In this approach, each of the fuzzy rules is successively temporarily removed, and the system error is checked to evaluate the system performance. The procedure of testing all rules in this manner allows removing the least beneficial rules, i.e. those which make the system error lowest after the rule removal. This means that a rule is removed only if it does not negatively impact the performance of the system. The rule reduction has been applied to each variant of the recommender (WM, WM+W, WM+D, WM+W+D).

Adding parameters to the system increases their degree of freedom but, at the same time, its complexity. Therefore, it is important to check how much the increase in the number of parameters improves system performance. In the proposed approach, it is worth checking whether the additional rule reduction allows significantly improving system performance. To evaluate the solutions, from this point of view, the Akaike Information Criterion (AIC) has been applied; see Sect. 3.2.3. The number of parameters, p, includes the system weights and the Dombi T-norm parameters for every rule. As the system error, Q, in the AIC, the RMSE is used.

3.3.3 Illustration of the Recommender Performance

In the simulations that illustrate the performance of the proposed systems, the MovieLens 10 M dataset has been used. For particular movies in the database, six attributes of the movies have been considered: *genre*, *year*, *keywords*, the same as in the recommenders A and B, and also *country*, *actors*, *directors*. In addition, the user rate of the movies, included in this database, have been applied as the decision attribute. From this dataset, 200 users that rated more than 30 movies have been selected.

Values of the *genre*, *country*, *actors*, *directors* have been coded according to Eq. (3.16).

For optimization of the weights and Dombi parameters, the evolutionary strategy $(\mu + \lambda)$ has been used, with the following parameters: (a) population size: 100, (b) number of iterations: 200, (c) crossover probability: 0.9, (d) mutation probability: 0.3, (e) mutation range: 0.2.

For the system evaluation (see Algorithm 2), k-fold cross-validation has been applied (with $k = 5$) where 80% of data samples have been used for learning and 20% for testing. It is worth noting that due to the proposed coding, the aggregated information about system outputs is partially incorporated in the encoding of inputs (also for testing data).

Simulation results concerning the RMSE (system error), for all users, are presented in Table 3.7, for each system, WM, WM+W, WM+D, and WM+W+D. Two versions of the recommenders are distinguished: with 3 and 6 inputs (3 and 6 attributes of the movies, respectively). The values of the RMSE are determined for learning and testing data. A comparison of the performance of these systems is illustrated in Fig. 3.15. We observe how the RMSE (denoted as rmse) depends on the percentage of rules reduced.

It is obvious, what we see in Table 3.7, that the average RMSE error of the recommendation systems with six inputs (corresponding to 6 attributes of the movies) has lower values than the RMSE of the systems with three inputs (only three attributes considered). Table 3.7 shows that the more attributes characterize the movies, the better recommendaitions will be (better performance of the recommender).

Figure 3.14 portrays isocriterial lines that represent constant values of the Akaike criterion (AIC) with different values of the system error, Q, and the number of parameters, p, for the systems under consideration. We see in Fig. 3.14 that the

Table 3.7 Average RMSE for all users; for 3 and 6 attributes (inputs)

System	Three inputs		Six inputs	
	Learning	Testing	Learning	Testing
WM	0.431	0.601	0.224	0.385
WM+W	0.329	0.562	0.167	0.361
WM+D	0.312	0.563	0.158	0.364
WM+W+D	0.312	0.562	0.152	0.359

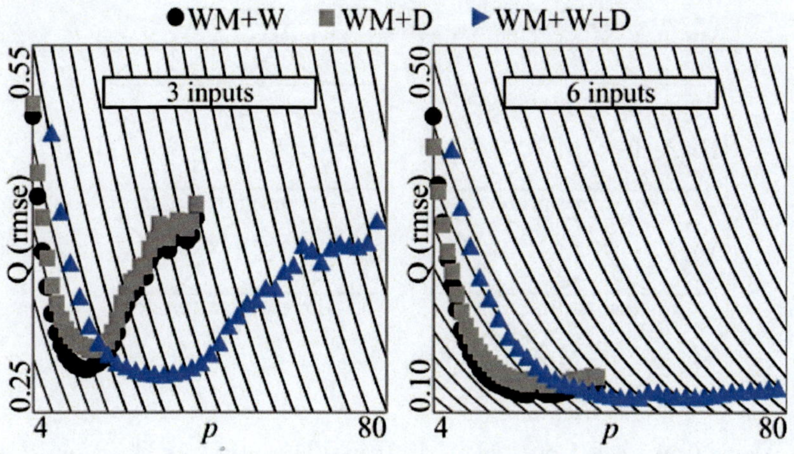

Fig. 3.14 Isocriterial lines representing the Akaike criterion for recommender C

optimal number of parameters should be low, for all considered systems (8–16 for the recommender with 3-inputs and 12–24 for 6-inputs).

Examples of fuzzy rules obtained for the fuzzy systems applied as recommender C are shown in Fig. 3.15. In addition, values of importance weights and the T-norm parameters, for particular rules, are presented. In this case (Fig. 3.15), also the recommender system with 6 inputs (6 attributes) is considered.

It is easy to present the fuzzy rules depicted in Fig. 3.15 in the explainable form, e.g., the first rule should be formulated as follows: IF x_1 is Medium AND x_2 is Very High AND x_3 is Very Low AND x_4 is Very Low AND x_5 is Medium AND x_6 is Medium THEN y is Medium.

Observing the RMSE error, we can also conclude that the additional reduction of the fuzzy rules improves the performance of the recommenders; for details, see [27]. Moreover, it is obvious that less number of rules results in better explanation facilities.

3.4 Conclusions Concerning Recommenders A, B, and C

The notations—recommender A, B, and C—have been introduced in order to distinguish three groups of the neuro-fuzzy recommendation systems. As the recommender A, the most popular neuro-fuzzy architecture, presented in Fig. 2.1, is employed and used during the learning of the system. This interpretable neuro-fuzzy network refers to the rules of the form (3.1). Consequent parts of these rules include crisp (constant) values, not fuzzy sets (or singletons that are fuzzy sets confined to single points—crisp values).

The second group, considered as the recommender B, applies the Mamdani fuzzy system and the rules of the form (2.1), with fuzzy sets in the consequent parts. Thus,

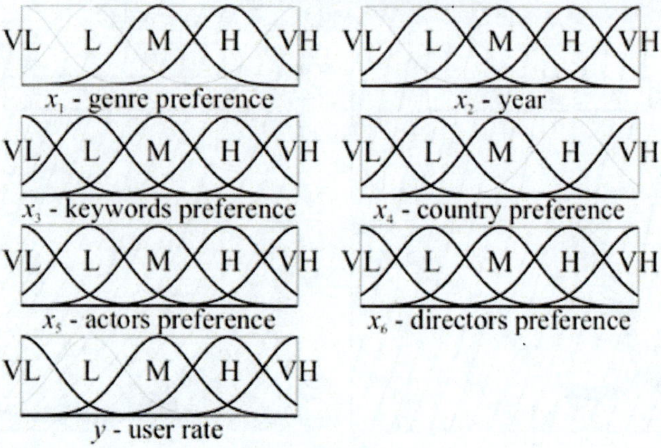

VL - Very Low, L - Low, M - Medium, H - High, VH - Very High

IF						THEN	w_j	p_j
x_1	x_2	x_3	x_4	x_5	x_6	y	=	=
M	VH	VL	VL	M	M	M	0.196	0.895
H	VH	H	L	M	M	M	0.625	0.575
M	VH	M	L	M	L	H	0.078	0.557
H	H	L	L	L	VL	VL	0.339	0.855
M	H	M	L	VL	L	VL	0.393	0.576
M	M	M	L	M	M	M	0.267	1.149
H	L	VH	M	VH	M	VH	0.515	0.750
M	M	L	L	VL	M	VL	0.979	0.689
M	VH	H	L	L	M	M	0.738	0.581
M	L	H	VH	H	H	H	0.633	1.806
H	M	M	M	L	M	M	0.140	1.565
H	VH	VH	L	VH	VH	VH	0.903	0.617

Fig. 3.15 Examples of fuzzy rules for recommender C

this group refers to a more general case of the neuro-fuzzy system but represented by the same connectionist architecture, illustrated in Fig. 2.1. However, although in both cases centers of the consequent fuzzy sets are parameters of the neuro-fuzzy network, the interpretation is different. In addition, the rule weights are also included in this connectionist network, and their values can be optimized by a learning algorithm.

Third group refers to the neuro-fuzzy systems that can be employed as the recommender C. This is a more general case that includes various modifications of the recommender B. This means that not necessary the Mamdani fuzzy system, where the product is commonly used as the T-norm, but also more sophisticated functions as the T-norm can be applied, e.g. the Dombi T-norm.

It should be emphasized that every kind of the connectionist neuro-fuzzy recommenders requires numerical values at their inputs, and produces numerical values at the output. Non-numerical (nominal) values of object attributes have to be transformed into corresponding numerical values. Different methods of encoding nominal values as their numerical counterparts, including the algorithms proposed for the particular recommenders described in this chapter, can be applied in any kind of the neuro-fuzzy recommendation system.

Different rule reduction procedures, as well as various optimization methods, e.g. evolutionary algorithms, can also be employed in any kind of the neuro-fuzzy recommenders.

The more complex is the neuro-fuzzy system, the more parameters include, e.g., rule weights, T-norm parameters. Therefore, the criteria presented in Sect. 3.2.3 for evaluating the balance between system accuracy and interpretability can also be used in every neuro-fuzzy recommender with many parameters.

Rule generation methods, including the WM, and NIT (as a matter of fact, both are very similar), as well as rule acquisition from experts, can be applied to various neuro-fuzzy recommenders.

Moreover, each of the recommendation systems A, B, and C, in application to the MovieLens data, can be considered with 3 or 6 inputs. The number of the recommender inputs equals to the number of object attributes. It is worth emphasizing that usually selection of appropriate attributes—the optimal number of the object features, and the most important ones—is a big challenge for specific problems.

With regard to recommenders A, B, and C, the difference concerns output values. As explained earlier, the recommender A is used only for the data where the decision attribute (rating value in the MovieLens dataset) takes a numerical value, e.g. 2, 3, 4, 5, and the system produces numerical values (not necessarily natural numbers) at the output. Recommenders belonging to groups B and C accept numerical and fuzzy output values. This means that values of the decision attribute can be viewed as fuzzy sets, for example, fuzzy numbers (about 2 or 3 or 4 or 5) that means approximately 2 or 3 or 4 or 5, with the interpretation by membership functions (e.g. Gaussian or triangular) with centers 2, 3, 4, 5, respectively.

In the case when fuzzy IF-THEN rules of the form (3.1) or (2.1) are known, maybe from experts, and a recommendation problem is viewed as a classification task, the neuro-fuzzy network portrayed in Fig. 2.2 can be employed for the recommenders A and B. Concerning the recommender C, similar neuro-fuzzy network can be used; however, it should be modified according to the T-norm applied in formula (3.18) instead of (2.8).

The RMSE is a popular measure of recommendation quality (see e.g., [7, 10]), by measuring the distance between predicted preferences and true preferences over items. It is worth noting that the RMSE was chosen for the Netflix Prize [2], to determine the accuracy of proposed recommenders for movie recommendations; based on the collaborative approach (see Chap. 1, Sect. 1.3) and the MovieLens datasets (available on-line: https://grouplens.org/datasets/movielens/).

Recommendation problems are usually viewed as classification tasks. However, as explained earlier, neuro-fuzzy recommenders can also be considered as systems for solving regression problems. It should be emphasized that the recommenders described in this chapter as classifiers concern multi-class classification, and as a special case—two-class problems (Yes/No).

Another kind of recommendation systems is applied in Chap. 4. This is a one-class classifier, created in the form of a neuro-fuzzy system (also using the WM). Having numerical data, the methods of features' encoding are not employed. However, in the case of nominal values of attributes, it is necessary to use a way of transformation into numerical values.

Concerning methods described in Sects. 3.1.1 and 3.3.1, presented in Fig. 3.1 and Eq. (3.16), respectively, it should be noted that the former is more intuitive although heuristic, while the latter is mathematically formulated. This means that the former method is better from an explainability point of view in spite of the fact that the latter can produce more accurate results.

In the next chapter, a recommendation system for investment advisers is constructed and applied for real data, and also two cases are distinguished: (1) with three attributes and (2) with more than twenty. For the multidimensional data, a special method of visualization of the data points is presented. The same method can also be used for data visualization when the multidimensional space of attributes is considered with regard to the multi-class recommenders. In addition, a new method of data visualization, more suitable for the recommender introduced in Chap. 4, is proposed and employed.

References

1. Alvarez-Estevez, D., Moret-Bonillo, V.: Revisiting the Wang–Mendel algorithm for fuzzy classification. Exp. Syst. **35**(4) (2018)
2. Bennett, J., Lanning, S.: The Netflix prize. In: Proceedings of KDD Cup and Workshop 2007. San Jose, California (2007)
3. Bozdogan, H.: Model selection and Akaike's information criterion (AIC): the general theory and its analytical extensions. Psychometrika **52**(3), 345–370 (1987)
4. Cpałka, K.: Design of Interpretable Fuzzy Systems. Studies in Computational Intelligence, vol. 684. Springer, Berlin (2017)
5. Dehghani, M., Riahi-Madvar, H., Hooshyaripor, F., Mosavi, A., Shamshirband, S., Zavadskas, E. K., Chau, K.: Prediction of hydropower geneation using Grey Wolf optimization adaptive neuro-fuzzy inference system. Energies **12**(2) (2019)
6. Dubois, D., Prade, H.: Fuzzy Sets and Systems: Theory and Applications. Academic, London (1980)
7. Ekstrand, M.D., Riedl, J.T., Konstan, J.A.: Collaborative filtering recommender systems. Found. Trends Human–Comput. Interact. **4**(2), 81–173 (2011)
8. Faris, H., Aljarah, I., Al-Betar, M.A., Mirjalili, S.: Grey wolf optimizer: a review of recent variants and applications. Neural Comput. Appl. **3**(2), 413–435 (2018)
9. Golafshani, E.M., Behnood, A., Arashpour, M.: Predicting the compressive strength of normal and hight-performance concretes using ANN and ANFIS hybridized Grey Wolf Optimizer. Constr. Build. Mater. **232**, 117266 (2020)
10. Gunawardana, A., Shani, G.: A survey of accuracy evaluation metrics of recommendation tasks. J. Mach. Learn. Res. **10**, 2935–2962 (2009)
11. Harper, F.M., Konstan, J.A.: The MovieLens datasets: history and context. ACM Trans. Interact. Intell. Syst. **5**(4), 19:1–19:19 (2015)

12. Ishibuchi, H., Nakashima, T.: Effect of rule weights in fuzzy rule-based classification systems. IEEE Trans. Fuzzy Syst. **9**(4), 506–515 (2001)
13. Ishibuchi, H., Yamamoto, T.: Rule weight specification in fuzzy rule-based classification systems. IEEE Trans. Fuzzy Syst. **13**(4), 428–435 (2005)
14. Jin, Y.: Fuzzy modeling of high-dimensional systems: complexity reduction and interpretability improvement. IEEE Trans. Fuzzy Syst. **8**(2), 212–221 (2000)
15. Kuncheva, L.: Fuzzy Classifier Design. Studies in Fuzziness and Soft Computing. Springer, Berlin (2000)
16. Mirjalili, S., Mirjalili, S.M., Lewis, A.: Grey Wolf optimizer. Adv. Eng. Softw. **69**, 46–61 (2014)
17. Nauck, D., Kruse, R.: How the learning of rule weights affects the interpretability of fuzzy systems. In: Proceedings of the IEEE International Conference on Fuzzy Systems 1998 (FUZZ-IEEE'98), vol. 2, pp. 1235–1240 (1998)
18. Nozaki, K., Ishibuchi, H., Tanaka, H.: A simple but powerful heuristic method for generating fuzzy rules from numerical data. Fuzzy Sets Syst. **86**(3), 251–270 (1997)
19. Prasad, M., Liu, Y.-T., Li, D.-L., Lin, C.-T., Shah, R.R., Kaiwartya, O.P.: A new mechanism for data visualization with TSK-type preprocessed collaborative fuzzy rule based system. J. Artif. Intell. Soft Comput. Res. **7**(1), 33–46 (2017)
20. Riid, A., Preden, J.-S.: Design of fuzzy rule-based classifiers through granulation and consolidation. J. Artif. Intell. Soft Comput. Res. **7**(2), 137–147 (2017)
21. Rutkowska, D.: Neuro-Fuzzy Architectures and Hybrid Learning. Physica-Verlag. Springer, Heidelberg, New York (2002)
22. Rutkowski, L.: Flexible Neuro-fuzzy Systems: Structures. Learning and Performance Evaluation. Kluwer Academic Publisher (2004)
23. Rutkowski, L.: Computational Intelligence: Methods and Techniques. Springer, Berlin (2008)
24. Rutkowski, T., Romanowski, J., Woldan, P., Staszewski, P., Nielek, R., Rutkowski, L.: A content-based recommendation system using neuro-fuzzy approach. In: 2018 IEEE International Conference on Fuzzy Systems (FUZZ-IEEE), pp. 1–8. IEEE (2018)
25. Rutkowski, T., Łapa, K., Nowicki, R., Nielek, R., Grzanek, K.: On explainable recommender system based on fuzzy rule generation techniques. In: Artificial Intelligence and Soft Computing, LNAI 11508, ICAISC 2019. Part I, pp. 358–372. Springer, Berlin (2019)
26. Rutkowski, T., Łapa, K., Nielek, R.: On explainable fuzzy recommenders and their performance evaluation. Int. J. Appl. Math. Comput. Sci. **29**(3), 595–610 (2019)
27. Rutkowski, T., Łapa, K., Jaworski, M., Nielek, R., Rutkowska, D.: On explainable flexible fuzzy recommender and its performance evaluation using the Akaike information criterion. In: International Conference on Neural Information Processing (ICONIP 2019), pp. 717–724. Springer, Berlin (2019)
28. Schwefel, H.-P: Evolution strategies: a family of non-linear optimization techniques based on imitating some principles of organic evolution. Ann. Oper. Res. **1**, 165–167 (1984)
29. Simiński, K.: Rule weights in a neuro-fuzzy system with a hierarchical domain partition. Int. J. Appl. Math. Comput. Sci. **20**(2), 337–347 (2010)
30. Söderström, T., Stoica, P.: System Identification. Prentice Hall International (1989)
31. Wang, L.-X., Mendel, J.M.: Generating fuzzy rules by learning from examples. IEEE Trans. Syst. Man Cybern. **22**(6), 1414–1427 (1992)

Chapter 4
Explainable Recommender for Investment Advisers

4.1 Introduction to the Real-Life Application of the Proposed Recommender

With the MovieLens examples, considered in Chap. 3, the way to recommend movies is to predict how would a user of the system rank a particular movie. If it is a movie that a user has not seen yet but would rank high, that is a potentially good recommendation. To be able to produce such a decision, the system needs to operate on a very specific training set (learning data) that includes the ratings for as many movies as possible. Based on those ratings and features (attributes) that describe the movies, the recommender system can predict the rating for new movies that the system has not known yet.

There is also an option, mentioned in Chap. 3, where objects are labeled Yes/No in a dataset, and based on that information, the system predicts which new objects to recommend or not to recommend. This is a special case of the multi-class recommendation problem and can be viewed as a binary (two classes) classification task.

Another situation concerns the case when the only available information is a list of objects that a user liked (no information about objects that users dislike). For example, we know which restaurants a user has visited over the past five years. He or she has not ranked those restaurants, so we have no other information. We do not know if he or she actually likes the restaurant. Nevertheless, it is an honest assumption to think that people usually visit restaurants that they like or they think that they would like. It is possible to figure out the taste of the user and recommend new restaurants based on this history.

Similar reasoning can be applied to analyzing financial markets strategies, especially with stock picking. We have only access to positive examples—things or actions that people have done. Hence, such problems can be treated as a one-class classification (see e.g. [11, 20, 21]). Based on some observations, it is possible to recommend new things (objects, actions).

T. Rutkowski, *Explainable Artificial Intelligence Based on Neuro-Fuzzy Modeling with Applications in Finance*, Studies in Computational Intelligence 964,
https://doi.org/10.1007/978-3-030-75521-8_4

One could argue that in the case of financial markets, stocks that were bought can mean "yes" and stocks that were sold mean "no". It is not that simple because investors sell for several different reasons, not only because they do not like the stock. Maybe they profited enough, and they want to move on, or they want to free some capital, or they are short-sellers, so they actually profit when stocks price decreases. But they buy only when they think that it fits their strategy. In this case, it is possible to analyze buy actions and recommend new stocks that are worth buying. The only thing that is needed, in order to produce a recommendation, is a log of transactions (the history—data of the previous transactions).

In the United States of America, every investment fund that manages over 100 million dollars has an obligation to submit a form with all transactions quarterly to SEC (Securities and Exchange Commission). The form is called 13F, and all data from this form are publicly available (see [2]). With all those transactions, it is possible to infer what to recommend next. The recommended stocks would not be the ones that will beat the market. It is not about the price, performance, or predicting returns per se. It is about fitting the strategy, the profile of preferences that is created based on past transactions. It can be seen as a filter that chooses stocks from the whole universe of available opportunities.

In this chapter, it is described how to design and implement an explainable neuro-fuzzy recommender system for stock recommendations based on information from 13F forms enriched with other data describing stocks.

4.2 Statement of the Problem

The problem is to recommend stocks that fit the strategy of investors. Now, the question is how to define a strategy—and the answer should refer to patterns from previous transactions.

As it was explained in Sect. 4.1, SEC Form 13F is a quarterly report that is filed by institutional investment managers with at least $100 million in equity assets under management. It discloses their U.S. equity holdings to the Securities and Exchange Commission (SEC) and provides insights into what the smart money is doing.

An example, illustrating the SEC Form 13F, is presented in Table 4.1 which includes actions of companies picked by one fund (Fund 1 in the Table) during the quarter that ends on December 31, 2018. Two transactions, "Add" and "New Buy", realized the companies, with regard to actions, have been considered, and their impact on the portfolio is indicated. In addition, values of minimum and maximum trading prices (TP), and change from the average percentage are shown.

Figure 4.1 portrays the process of generating recommendations based on the 13F form that is a part of the dataset prepared for the recommender. The dataset also includes values of attributes that characterize the companies. It should be emphasized that there are only "Add" and "New Buy" in column "action" in Table 4.1. This means that only positive labels are taken into account. Why? Because the reasons for "Reduce" and "Sold Out" actions are not known, so for those cases, the system cannot

Table 4.1 An example of the SEC Form 13F

Ticker	Picked By	Portfolio date	Action	Impact to portfolio	Min TP	Max TP	Chaged from Avrg (%)	Commment	% of shares Outstanding
AMZN	Fund 1	2018-12-31	Add	3,37	1343,96	1971,31	5,87	Add 64.88%	0,03
LTRPB	Fund 1	2018-12-31	New Buy	2,62	18,93	18,93	−48,49	New holding	4,97
LEXEA	Fund 1	2018-12-31	Add	2,49	37,46	46,42	21,7	Add 334.06%	3,69
BkNG	Fund 1	2018-12-31	Add	1,45	1616,83	1998,17	3,68	Add 912.85%	0,05
REZI	Fund 1	2018-12-31	New Buy	1,39	19,35	31	−39,77	New holding	1,4
DISCK	Fund 1	2018-12-31	Add	1,34	22,82	30,62	−8,81	Add 137.69%	0,49
WCG	Fund 1	2018-12-31	New Buy	1,27	221,21	319,24	−0,9	New holding	0,27
GOOGL	Fund 1	2018-12-31	Add	1	984,67	1211,53	6,91	Add 79.28%	0,01
CFFAU	Fund 1	2018-12-31	New Buy	0,94	10	10	5	New holding	–
HTA	Fund 1	2018-12-31	New Buy	0,91	24,48	29,09	4,53	New holding	0,45
UAL	Fund 1	2018-12-31	Add	0,59	78,93	96,7	−6,22	Add 982.84%	0,07

infer the decision—do not recommend. Therefore, the recommendation problem is treated as a one-class classification, as explained in Sect. 4.1.

An illustration of the real data applied in the recommender, from the companies shown in Table 4.1, is presented in Tables 4.2, 4.3 and 4.4. As a matter of fact, Tables 4.2, 4.3 and 4.4 presents a combination of data coming from two sources: the history of transactions from the 13F and values of attributes characterizing the companies. Both sources of the data (Form 13F and "Trading Journal") are indicated in the first block of the scheme presented in Fig. 4.1. More information concerning the data are included in Sect. 4.3.

The second block, in Fig. 4.1, data enrichment, denotes the process of the dataset's preparation using the data from two sources in the first block. This is described in Sect. 4.3.1. By data enrichment, we can understand merging data from an external source with an existing database.

The third block—features' selection and preprocessing—also refers to Sect. 4.3, especially to Sect. 4.3.2 and Appendix A, where attributes characterizing, companies are described. The data, including values of those attributes, are employed in order to determine new recommendations by the recommender. Visualizations of the data are presented in Sect. 4.3.3. Thus, the first (upper) part of the scheme concerns the dataset preparation for analysis and inference by the recommender.

Now, let us describe how the recommendations are produced by the one-class neuro-fuzzy classifier (recommender) based on the data prepared from the 13F and characteristics (attributes) of the companies. However, at first, the next three blocks

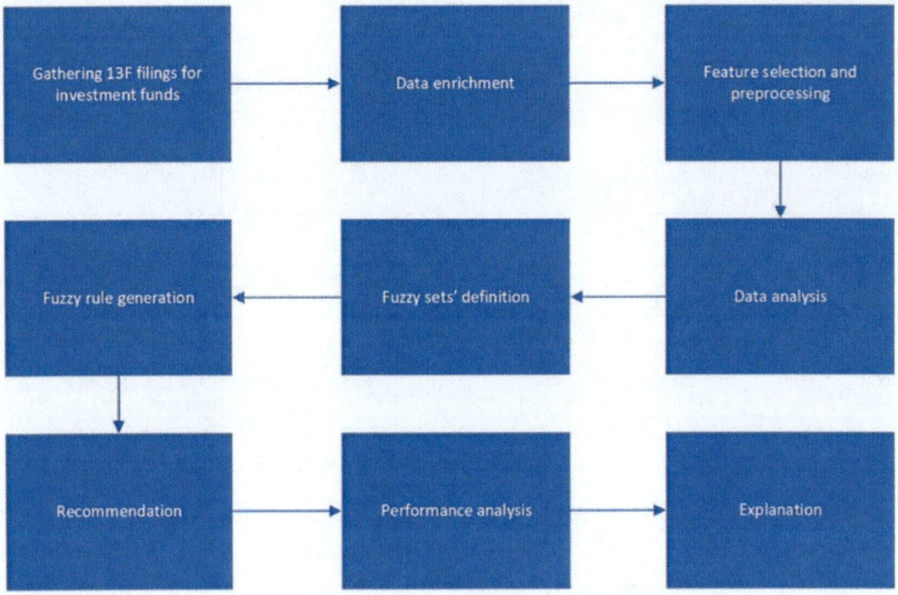

Fig. 4.1 Scheme of generating and explaining recommendations for investment advisers

(in the middle part of the scheme) should be considered: data analysis, fuzzy sets' definition, and fuzzy rules' generation. This part is very important because, based on the data analysis, fuzzy sets are defined, and then fuzzy IF-THEN rules are generated. The recommender infers decisions concerning recommendations based on the fuzzy rules. Moreover, the fuzzy sets and rules are applied in order to explain the decisions (recommendations). Thus, let us focus our attention on the middle part of the scheme.

To analyze the data, with regard to their distribution in the domains of particular attributes, histograms are determined, and then—fuzzy intervals and fuzzy sets that are used in fuzzy IF-THEN rules. This is presented in Sect. 4.4, where the fuzzy sets' definition concerning the data analysis (by an expert) is described, so the fuzzy sets reflect values of the features (attributes). Different shapes of membership functions of the fuzzy sets are proposed, and then applied in the fuzzy IF-THEN rules.

The fuzzy rules' generation, as the last block in the middle part of the scheme, refers to Sect. 4.5. Two kinds of fuzzy IF-THEN rules are determined: the recommendation rules and explanation rules. As the names indicate, the former is applied in order to produce recommendations, while the latter is employed for an explanation of the recommender's decisions.

The last part of the scheme (the lower part) refers to the recommender's performance that is described in Sect. 4.6. As mentioned earlier, the recommender infers decisions based on the recommendation rules that are formulated in Sect. 4.6.1. Visualizations of the results are illustrated in Sect. 4.6.2. Explanations of the recommendations are presented in Sect. 4.6.3.

The same process of generating and explaining the recommendations for investment advisers, as portrayed in Fig. 4.1, with more details, is illustrated in Fig. 4.2. The first and second kinds of rules refer to the explanation rules and recommendation rules, respectively. The recommendation rules are based on the fuzzy points representing past transactions, while the explanation rules use fuzzy sets defined based on the fuzzy intervals; see Sects. 4.4, 4.5, and 4.6.

4.3 Description of the Datasets and Feature Selection

Now, the data enrichment (see Sect. 4.3.1) as well as features' description and selection (see Sects. 4.3.2 and Appendix A) are presented. In addition, in Sect. 4.3.3, a method for multidimensional data visualization is illustrated.

4.3.1 Data Enrichment and Dataset Preparation

As mentioned in Sect. 4.2, the dataset is obtained from two sources: SEC Form 13F that is a quarterly report of the transactions and companies characteristics from "Trading Journal". Combining both sources of the information, the data for the recommender is prepared in the form presented in Tables 4.2, 4.3 and 4.4. In this way, we have got datasets for particular funds; let us notice that Tables 4.2, 4.3 and 4.4 contain data for one fund (Fund1). All the data items include attribute values with decision labels referring to only one class ("Add" and "New Buy" mean "to recommend").

The datasets for particular funds allow discovering patterns that correspond to strategies of investments realized by those funds. Based on these data, the recommender system produces recommendations for investment advisors according to the individual strategies.

Tables 4.2, 4.3 and 4.4 illustrate real data from the selected asset management companies. The attributes that characterize the companies are described in Sect. 4.3.2 and Appendix A.

4.3.2 Description of Selected Attributes—Simplified Version

Twenty one attributes are considered in the recommendation problem. Two versions are distinguished: full and simplified. The former includes all twenty one attributes while the latter only three selected ones. All of them characterize the companies (see Tables 4.2, 4.3 and 4.4). The description of the features comes from (www.investopedia.com).

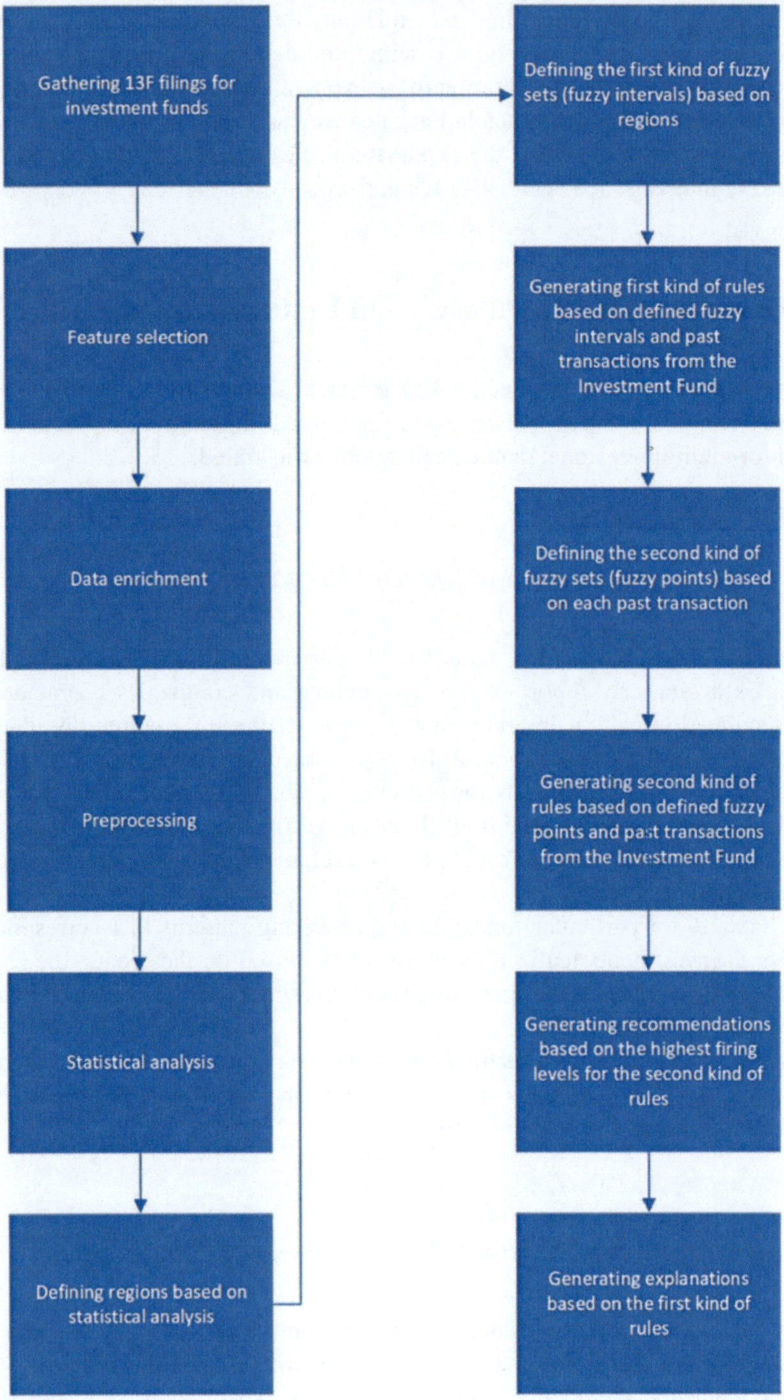

Fig. 4.2 More detailed scheme of generating and explaining recommendations for investment advisers

Table 4.2 Fragment of real data from asset management companies: part 1

Ticker	Picked By	Action	Date	Fiscal period	Company name	Currentration	Eps growth	Revenue growth
AMZN	Fund 1	Add	12/31/2018 12:00:00AM	Dec 18	AMZN—Amazon.com Inc	1,1	61,07	18,78
LTRPB	Fund 1	New Buy	12/31/2018 12:00:00AM	Dec 18	LTRPB—Liberty TripAdvisor Holding Inc	1, 52	97,76	13,86
LEXEA	Fund 1	Add	12/31/2018 12:00:00AM	Dec 18	LEXEA-Liberty Expedia Holdings Inc	0,66	83,83	9,37
BkNG	Fund 1	Add	12/31/2018 12:00:00AM	Dec 18	BKNG—Booking Holding Inc	2,36	221,47	22,37
REZI	Fund 1	New Buy	12/31/2018 12:00:00AM	Dec 18	REZI—Resideo Technologies Inc	1,21	0	10,75
DISCK	Fund 1	Add	12/31/2018 12:00:00AM	Dec 18	DISCK—Discovery Inc	1,06	119,1	18,24
WCG	Fund 1	New Buy	12/31/2018 12:00:00AM	Dec 18	WellCare Health Plans Inc	1,36	−17,16	24,74
GOOGL	Fund 1	Add	12/31/2018 12:00:00AM	Dec 18	GOOGL—Alphabet Inc	3,92	393,56	22,24
HTA	Fund 1	New Buy	12/31/2018 12:00:00AM	Dec 18	HTA—Healthcare Trust of America Inc.	1,72	−65	−1,44
UAL	Fund 1	Add	12/31/2018 12:00:00AM	Dec 18	UAL—United Continental Holdings Inc.	0,54	−14,57	18,65

Table 4.3 Fragment of real data from asset management companies: part 2

Ticker	Total current assets	Profit margin	Debt to equity	Dividend yield	Ebitda	Evtoebitda	52 weeklow	52 weeklow	Free cash flow
AMZN	75101	4,18	0,914	0	7999	26,27	2004,36	1343,96	12743
LTRPB	932	−2,31	1,449	0	51	18,56	23,91	14,61	14
LEXEA	5304	−0,58	1,767	0	409	15,68	46,92	37,46	−401
BKNG	8407	20,1	0,985	0	806,267	14,86	1998,17	1616,83	997,823
REZI	1809	1,26	0,783	0	122	8,77	31	19,35	69
DISCK	4231	9,72	2,033	0	1827	4,77	31,53	22,83	888
WCG	6832,4	0,92	0,502	0	181,5	9,84	321,88	221,21	15
GOOGL	135676	22,78	0,023	0	12681	14,11	1211,53	984,67	5867
HTA	319,054	8,9	0,78	0	109,628	13,05	29,48	24,48	79,911
UAL	7194	4,4	1,474	0	726	10,17	96,7	78,93	−484

Table 4.4 Fragment of real data from asset management companies: part 3

Ticker	Total gross profit	Operating margin	Divpayoutratin	Price to book	Price to earnings	Roa	Roe	Debt	Total debt to current asset
AMZN	17569	5,23	0	16,93	1,62	7,9	29,29	68391	0,245
LTRPB	259	2,6	0	4	0	−0,6	−9,38	613	0,093
LEXEA	2117	−3,3	0	0,93	0	−0,2	−2,48	8078	0,133
BKNG	3212,615	35,58	0	9,08	1,14	11	27,58	3555	0,381
REZI	330	8,29	0	1,65	0	1,32	3,03	1489	0,242
DISCK	1863	28,8	0	1,44	17,03	3,34	6,98	3997	0,524
WCG	720,9	2,05	0	2,78	0,99	1,85	5,32	5016	0,181
GOOGL	21358	20,89	0	4,09	1,55	15,8	20,6	34620	0,017
HTA	116,951	22,34	4,429	1,64	3,53	0,98	1,86	185, 1	0,411
UAL	2857	9,02	0	2,28	0,33	4,12	18,99	13212	0,329

The following attributes have been selected for the simplified version of the recommendation problem:

currentratio:
The current ratio is a liquidity ratio that measures a company's ability to pay short-term obligations or those due within one year. It tells investors and analysts how a company can maximize the current assets on its balance sheet to satisfy its current debt and other payables.

evtoebitda:
The enterprise-value-to-EBITDA ratio varies by industry. The EV/EBITDA for the S&P 500 has typically averaged from 11 to 14 over the last few years. As of June 2018, the average EV/EBITDA for the S&P was 12.98. As a general guideline, an EV/EBITDA value below 10 is commonly interpreted as healthy and above average by analysts and investors.

pricetobook:
Companies use the price-to-book ratio to compare a firm's market to book value by dividing price per share by Book Value Per Share (BVPS). An asset's book value is equal to its carrying value on the balance sheet, and companies calculate it netting the asset against its accumulated depreciation.

4.3.3 Multidimensional Data Visualization

The datasets from asset management companies include many attributes. Thus, the data items are represented as points in a multidimensional space. In order to visualize the data and be able to observe hidden patterns, a method of dimensionality reduction should be applied. For this purpose, the most suitable is the t-SNE visualization technique. This is a method for exploring high-dimensional data by giving each data point a location in a two or three-dimensional map [22]. This is a variation of the SNE (Stochastic Neighbor Embedding) introduced in [9].

The t-SNE method preserves the neighborhood of the points. This means that the data points that are located close to each other in the 3D space are grouped as

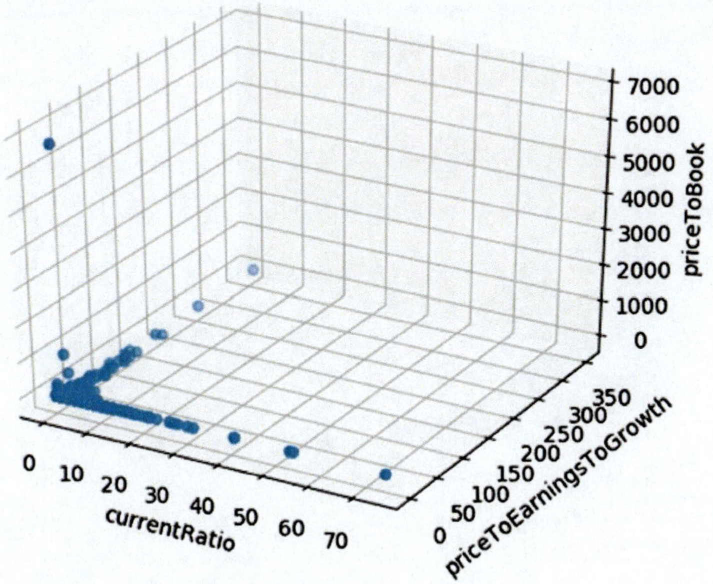

Fig. 4.3 Visualization of the real data in 3D attribute space

the neighbors somewhere in the 2D space; however, distances between the groups (clusters) are not preserved.

Since each real object is represented as one point in a feature space, thus the data points located near in this (multidimensional) space characterize real objects that are similar with regard to the attributes of this space. Hence, a neighborhood of the data points represents similar objects in the original multidimensional space, as well as in the space obtained after the t-SNE transformation. Of course, from this point of view, it does not matter how far the clusters of similar data points are located in the space transformed by the t-SNE method.

A tuneable parameter, "perplexity", says (loosely) how to balance attention between local and global aspects of the data. The perplexity can be interpreted as a smooth measure of the effective number of neighbors. Typical values of the perplexity are between 5 and 50. The perplexity should be smaller than the number of data points.

Visualizations of the real data items, considered in this chapter, are presented in the 3D and 2D spaces of attributes, in the case of three attributes from those described in Sect. 4.3.2 and Appendix A, when the t-SNE method is not applied. Figure 4.3 portrays the scatter plot for every data points in the attribute domains. Figure 4.4 shows the data items within smaller ranges, where more values of the data attributes are gathered; the data items located at the ends of the attribute domains are removed (as outliers).

Figure 4.5 presents the 2D scatter plot for two of the three attributes (currentratio and pricetobook) for every data items, while the version with the outliers removed is

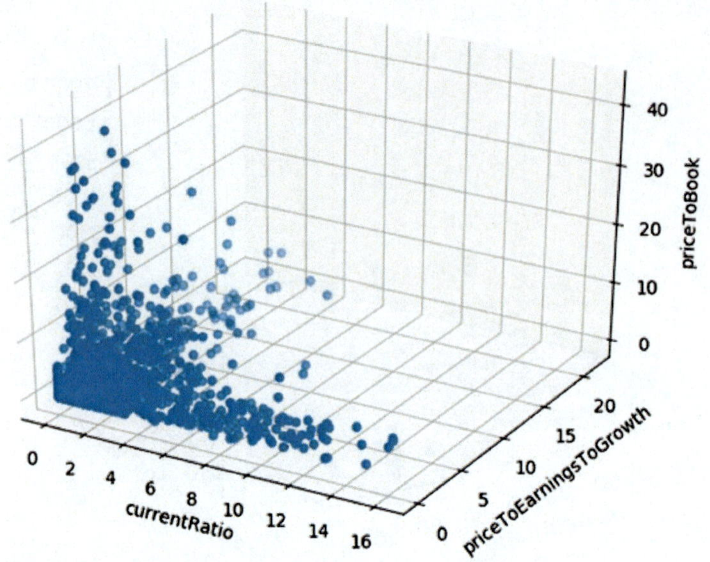

Fig. 4.4 Visualization of the real data in the 3D space (outliers removed)

depicted in Fig. 4.6. Similar visualizations of these data items, but in the 2D spaces of different two attributes, are portrayed in the next figures. Thus, Figs. 4.7 and 4.8 illustrate the 2D scatter plots for attributes: currentratio and pricetoearnings; the full version and with outliers removed, respectively. Analogously, Figs. 4.9 and 4.10 portray the 2D scatter plots for attributes: pricetoearnings and pricetobook; also the full version and with outliers removed, respectively.

Figures 4.11, 4.12 and 4.13 illustrate the visualizations of the real data in the 21-dimensional space when the t-SNE method is used. Figure 4.11 shows the data point in the 3D space, while Figs. 4.12 and 4.13 portray the data in the 2D space, with different values of the perplexity: 5 and 100, respectively.

As we see, and what is obvious, when only three attributes are considered, it is possible to visualize the data points directly in the 3D attribute space or pairwise in the 2D spaces. However, when more attributes are included, the t-SNE method is useful for data visualization.

4.4 Definition of Fuzzy Sets

With regard to the problem formulated in Sect. 4.2, and the attributes described in Sect. 4.3, the fuzzy approach is proposed. This means that values of the attributes are viewed as fuzzy sets [23], defined by membership functions. The recommender created for this problem, first of all is a fuzzy system [24] that infers based on fuzzy IF-THEN rules.

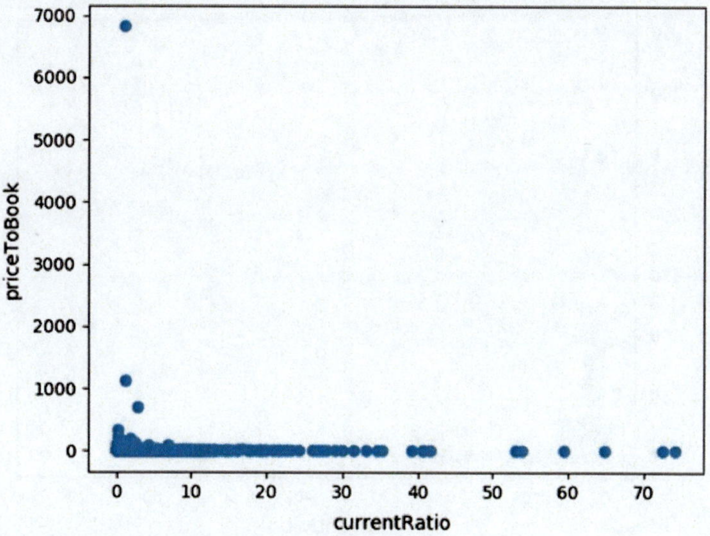

Fig. 4.5 Visualization of the real data in 2D attribute space

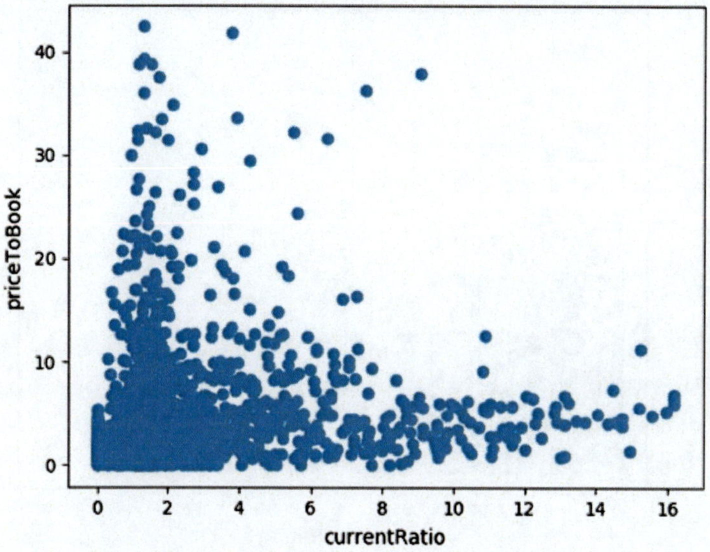

Fig. 4.6 Visualization of the real data in the 2D space (outliers removed)

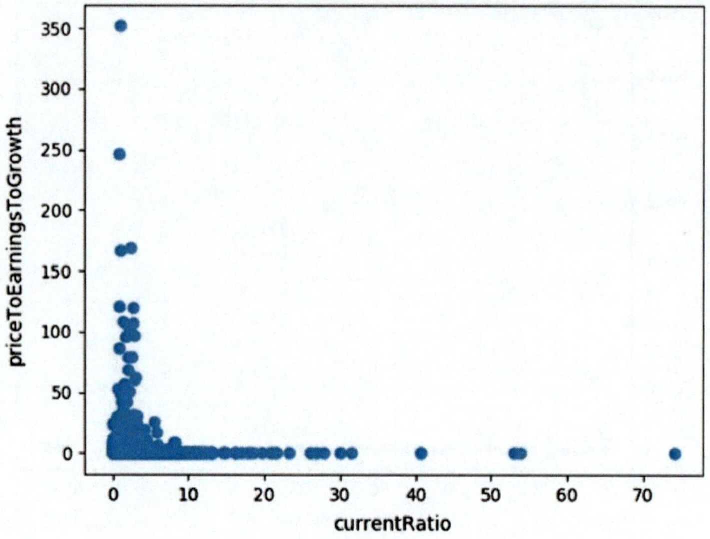

Fig. 4.7 Visualization of the real data in 2D attribute space; different attributes

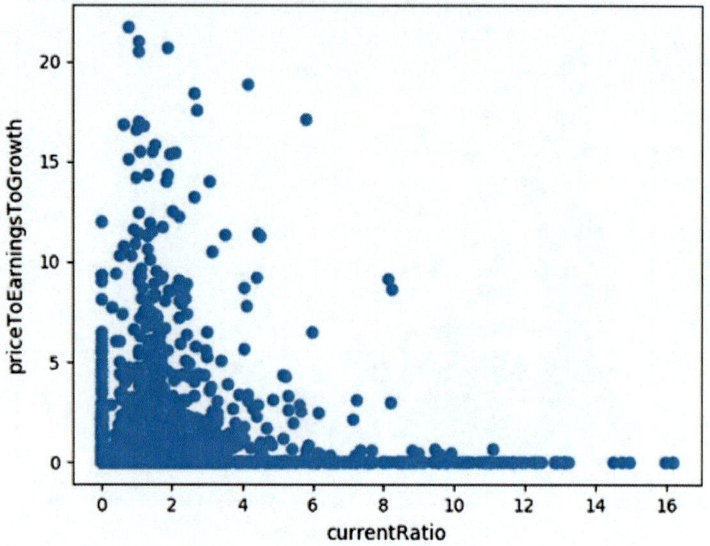

Fig. 4.8 Visualization of the real data in the 2D space of different attributes (outliers removed)

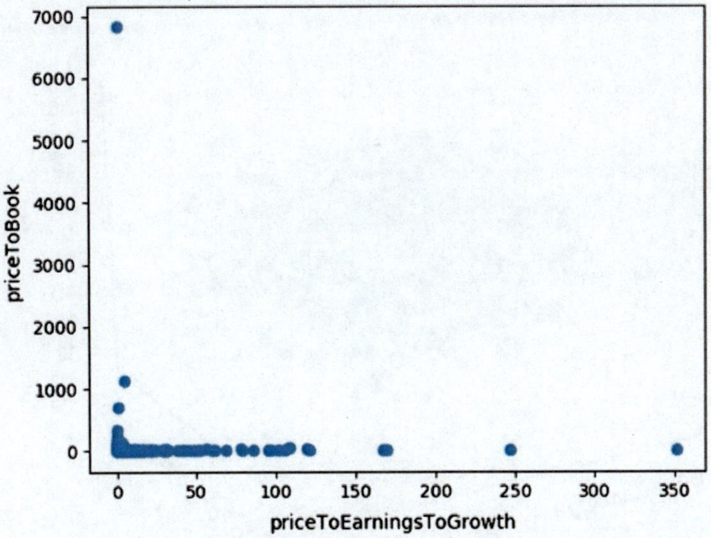

Fig. 4.9 Visualization of the real data in 2D attribute space; another pair of attributes

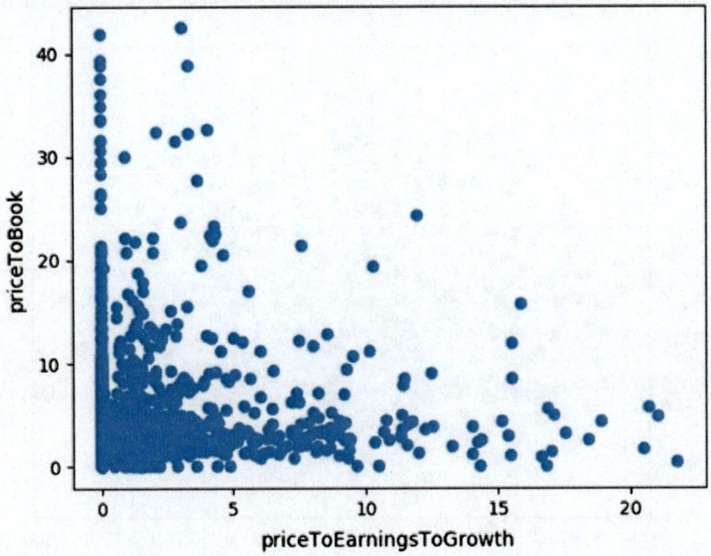

Fig. 4.10 Visualization of the real data in the 2D space of another pair of attributes (outliers removed)

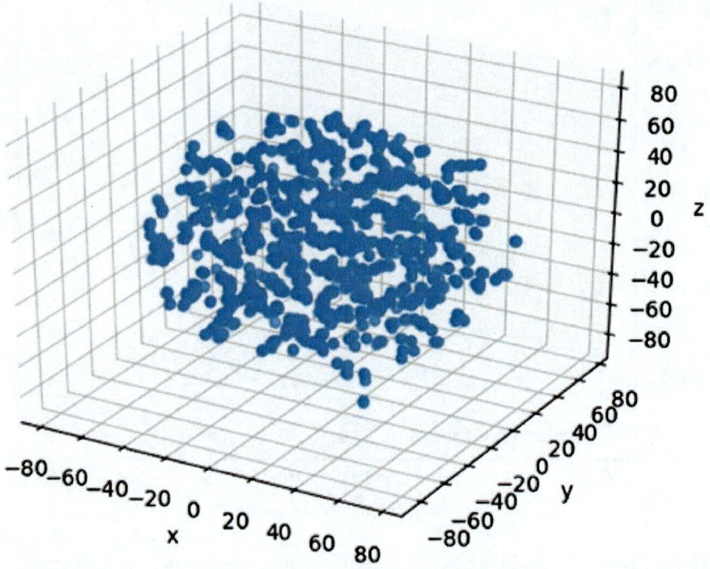

Fig. 4.11 Visualization of the real data using t-SNE method—transactions from Fund 1 (3D)

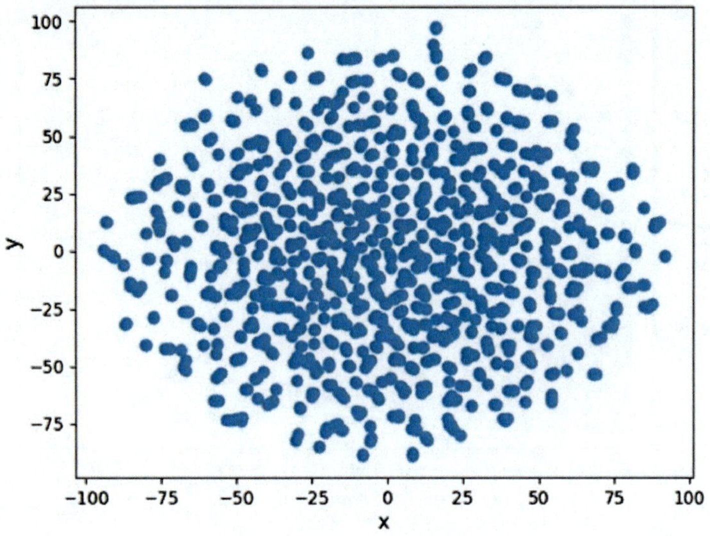

Fig. 4.12 Visualization of the real data using t-SNE method—transactions from Fund 1 (2D, perplexity 5)

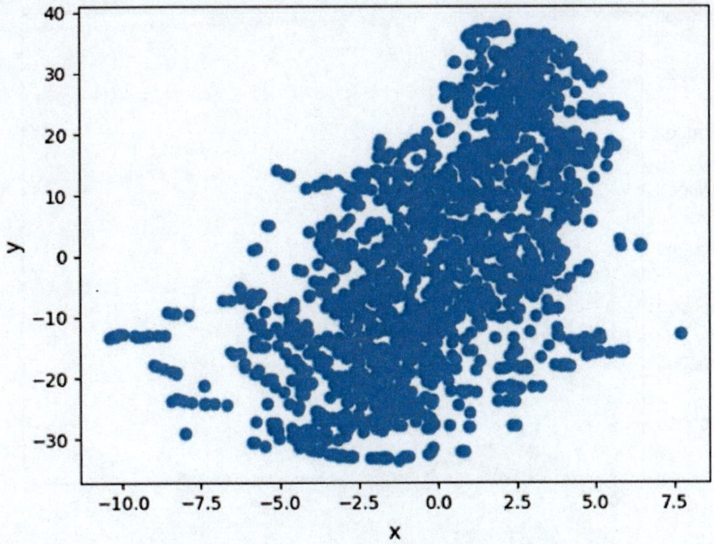

Fig. 4.13 Visualization of the real data using t-SNE method—transactions from Fund 1 (2D, perplexity 100)

For every attribute, described in Sect. 4.3.2 and Appendix A, fuzzy sets with linguistic labels have been defined by use the trapezoidal membership functions:

$$\mu_{A_i}(x_i; a, b, c, d) = \max\left(\min\left(\frac{x_i - a}{b - a}, 1, \frac{d - x_i}{d - c}\right), 0\right) \tag{4.1}$$

and z-shaped membership functions:

$$\mu_{A_i}(x_i; \alpha, \beta) = \begin{cases} 1 & \text{for } x_i \leq \alpha \\ 1 - 2\left((x_i - \alpha)/(\beta - \alpha)\right)^2 & \text{for } \alpha \leq x_i \leq (\alpha + \beta)/2 \\ 2\left((x_i - \beta)(\beta - \alpha)\right)^2 & \text{for } (\alpha + \beta)/2 \leq x_i \leq \beta \\ 0 & \text{for } x_i \geq \beta \end{cases} \tag{4.2}$$

as well as s-shaped membership functions:

$$\mu_{A_i}(x_i; \gamma, \delta, \eta) = \begin{cases} 0 & \text{for } x_i \leq \gamma \\ 2\left((x_i - \gamma)(\eta - \gamma)\right)^2 & \text{for } \gamma \leq x_i \leq \delta \\ 1 - 2\left((x_i - \eta)/(\eta - \gamma)\right)^2 & \text{for } \delta \leq x_i \leq \eta \\ 1 & \text{for } x_i \geq \eta \end{cases} \tag{4.3}$$

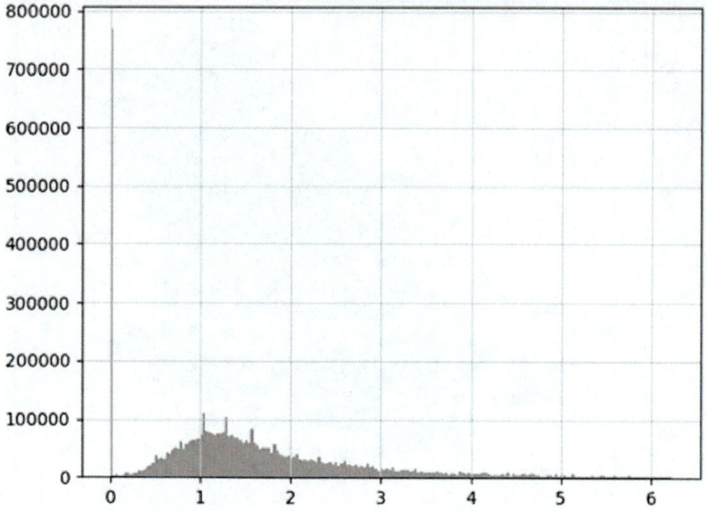

Fig. 4.14 Histogram of attribute: currentratio

where $a, b, c, d, \alpha, \beta, \gamma, \delta, \eta$ are parameters of the membership functions, different for each linguistic variable x_i, for $i = 1, 2, \ldots, n$, where n denotes the number of attributes. These parameters also differ for particular membership functions computed within particular domains of the attributes.

Semantic meanings of the fuzzy sets are: Very Low, Low, Medium, High, Very High. Of course, the meaning of these labels is different for particular linguistic variables, corresponding to the attributes, and depend on their different domains.

The fuzzy sets (membership functions) are defined based on the data analysis, according to formulas (4.1), (4.2), and (4.3). Figures 4.14, 4.18, and 4.22 portray histograms of the data for three attributes selected in the simplified version (see Sect. 4.3.2), i.e.: currentratio, evtoebitda, and pricetobook. The points of 5, 25, 50, 75 and 95th percentile have been determined (see Figs. 4.15, 4.19, and 4.23) in order to construct the fuzzy sets in Figs. 4.16, 4.20, and 4.24, respectively. The total number of the observed attributes' values is equal to 5 millions. For example, if the attribute currentRatio is considered, then 95th percentile of the observations corresponds to the value 6.2 as it is indicated in Fig. 4.15.

The WM (Wang-Mendel) method of rule generation, described in Chap. 2, Sect. 2.4, and used in Chap. 3, along with the fuzzy sets defined based on the histograms depicted in Figs. 4.14, 4.18, and 4.22, allowed generating fuzzy IF-THEN rules presented in Sect. 4.5. For the full version of the attributes (see Appendix A), in the similar way, the fuzzy membership functions have been defined, in Appendix E, based on the additional histograms, presented in Appendix D, and the fuzzy IF-THEN rules included in Appendix C have been formulated.

Apart from the membership functions portrayed in Figs. 4.16, 4.20, and 4.24, and those in Appendix E, the fuzziness to every data point in the dataset is introduced. This means that the data points representing the past transactions (investments) of a partic-

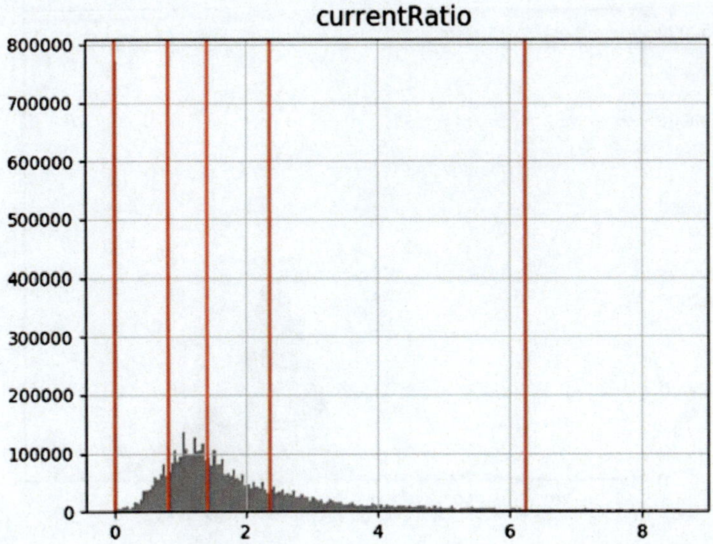

Fig. 4.15 Reference points for regions of attribute: currentratio (5, 25, 50, 75 and 95th percentile)

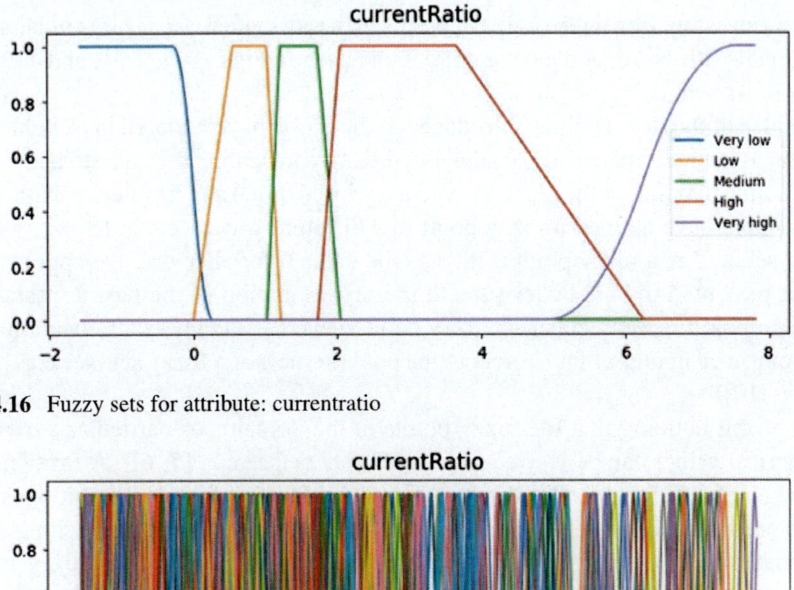

Fig. 4.16 Fuzzy sets for attribute: currentratio

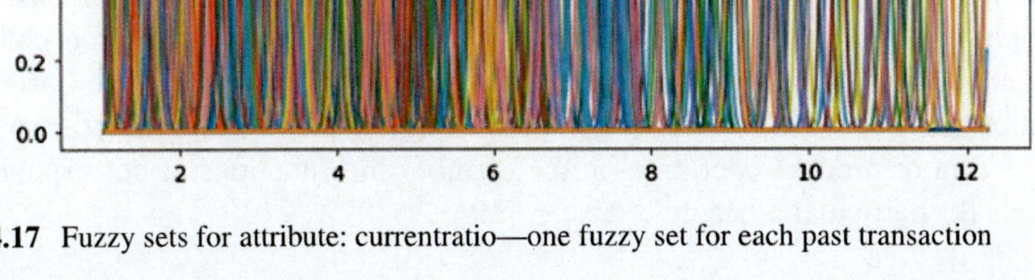

Fig. 4.17 Fuzzy sets for attribute: currentratio—one fuzzy set for each past transaction

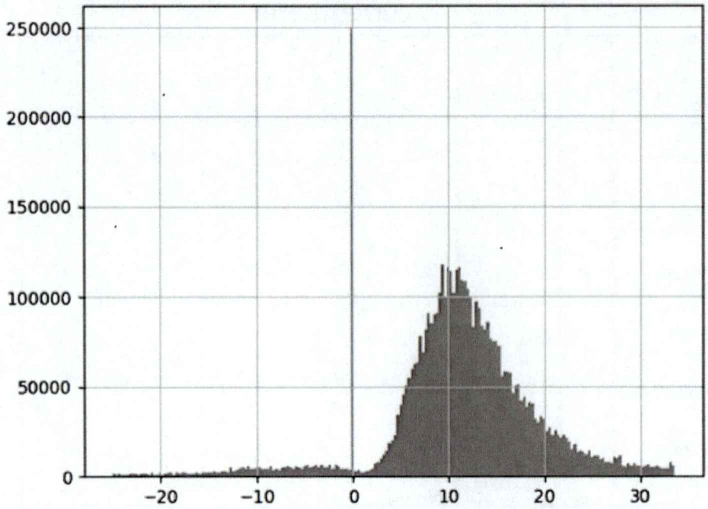

Fig. 4.18 Histogram of attribute: evtoebitda

ular user (investor) are viewed as fuzzy points in the attribute space. This is illustrated, for particular attributes, and an example of one user, in Figs. 4.17, 4.21, and 4.25, as well as in Appendix F.

The idea of the fuzzy points, introduced to the problem considered in this chapter, is similar as in [12], where a fuzzy point is a region representing the uncertain location of an Euclidean point; in the plane viewed as a closed circle and its interior. Although other authors have defined a fuzzy point in a different way (see e.g. [3, 8]), where a fuzzy set in X is a fuzzy point if it takes the value 0 for all $x \in X$ except one, the concept presented in [12] better suits to the representation of the past transactions for the proposed recommender. A fuzzy point viewed as the uncertain location of a crisp point, as a matter of fact, refers to the core (kernel) of a fuzzy set (see e.g. [19], and also [10]).

It is worth noticing that the fuzzy points in the domains of particular attributes of numerical values can be viewed as fuzzy numbers (see e.g. [5, 6]). A fuzzy number is a generalization of a regular (crisp) real number in the sense that it does not refer to one single value but rather to a connected set of possible values, with grades (of membership) between 0 and 1. In a multidimensional space, the fuzzy numbers (fuzzy points) constitute a fuzzy vector that is a multidimensional fuzzy point (see [12]).

Usually, fuzzy points (understood as explained above) in the multidimensional space of attributes are determined as the Cartesian product of the fuzzy sets characterizing this point in particular attribute domains (one-dimensional). The Cartesian product of fuzzy sets is also a fuzzy set, defined by the membership function determined as min or product operation on the membership functions of corresponding fuzzy sets for particular attributes; see e.g. [19].

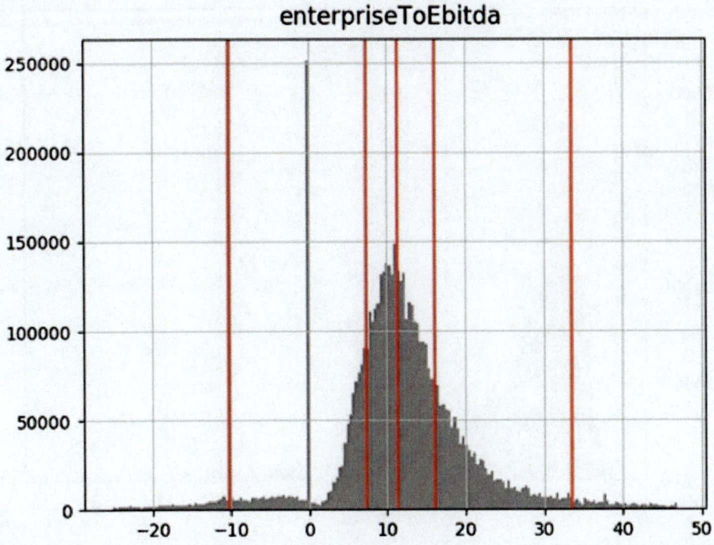

Fig. 4.19 Reference points for regions of attribute: evtoebitda (5, 25, 50, 75 and 95th percentile)

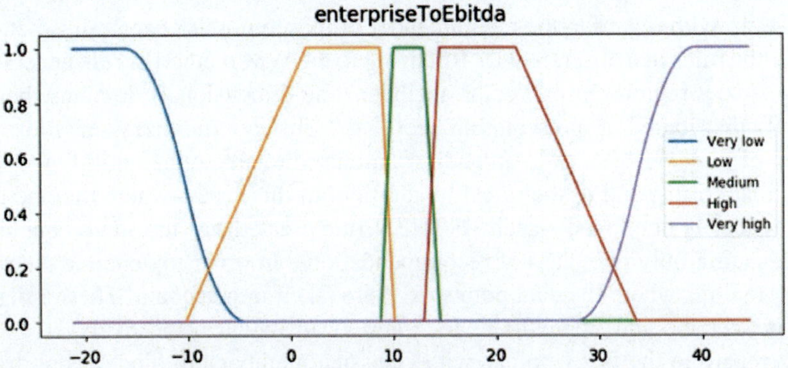

Fig. 4.20 Fuzzy sets for attribute: evtoebitda

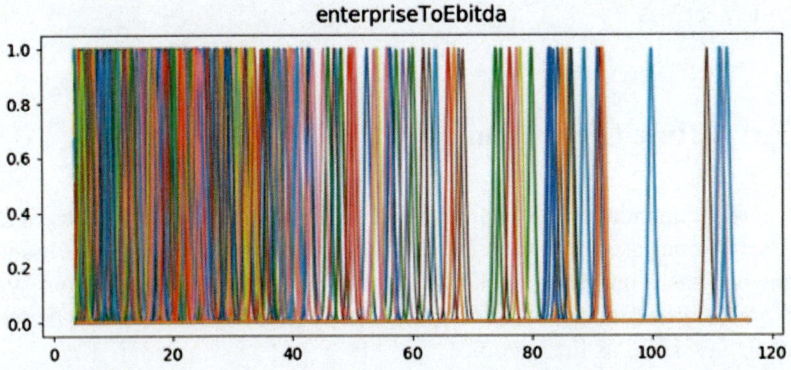

Fig. 4.21 Fuzzy sets for attribute: evtoebitda—one fuzzy set for each past transaction

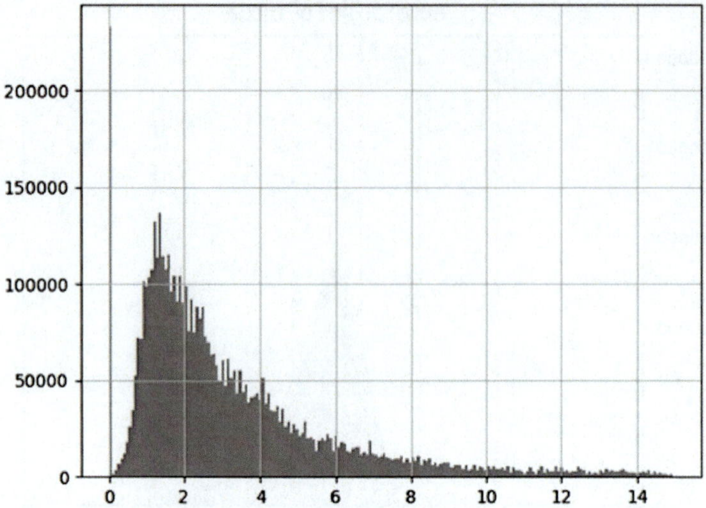

Fig. 4.22 Histogram of attribute: pricetobook

However, with regard to the recommender proposed in this chapter, it is better to consider the rules in a different way. Referring to the WM method of rule generation, we have fuzzy regions in particular attribute (one-dimensional) domains, but not uniformly distributed, as presented in Sect. 2.4.2. Instead, the fuzzy sets illustrated in Figs. 4.16, 4.20, 4.24, and Appendix E, are used to create the multidimensional regions that correspond to the rules. In addition, in this case—where the one-class recommender is employed—each IF-THEN rule created has the same conclusion part, indicating only one class of recommendations. In order to generate the rules, only the regions where the data points exist are taken into account. These rules are shown in Sect. 4.5 and Appendix C, for 3 and 21 attributes, respectively.

With regard to the fuzzy points, with Gaussian membership functions portrayed in Figs. 4.17, 4.21, 4.25, and Appendix F, fuzzy IF-THEN rules have been created for each data point of the past transactions. An example of 5 rules of this kind is also included in Sect. 4.5.

4.5 Fuzzy Rule Generation

The novel recommendation system presented in this chapter uses two different kinds of fuzzy sets—one based on a statistical analysis of all transactions from thousands of funds and one based on transactions from the fund that the recommendations system is built for. The first kind are based on fuzzy sets described in Sect. 4.4). The second kind can be described as fuzzy points.

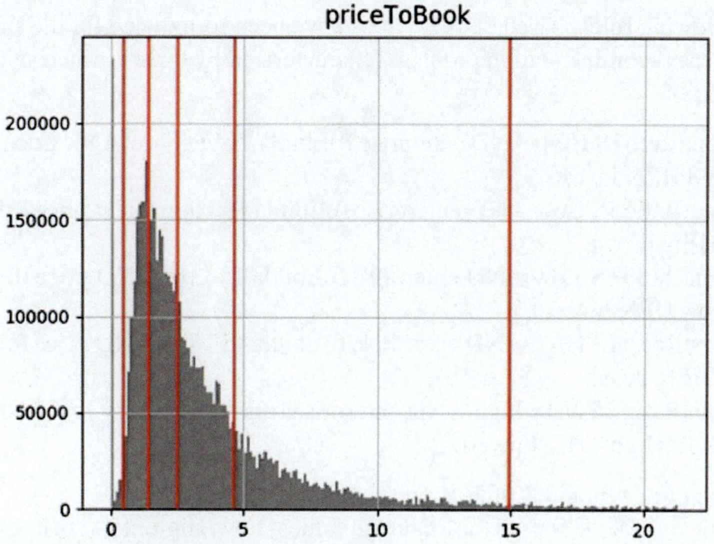

Fig. 4.23 Reference points for regions of attribute: pricetobook (5, 25, 50, 75 and 95th percentile)

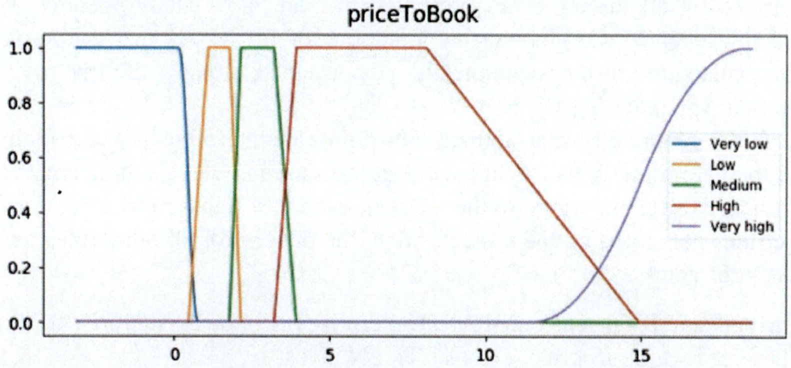

Fig. 4.24 Fuzzy sets for attribute: pricetobook

Fig. 4.25 Fuzzy sets for attribute: pricetobook—one fuzzy set for each past transaction

The following rules, based on fuzzy sets, have been formulated for the simplified version of the recommendation problem (characterized by three attributes presented in Sect. 4.3.2):

1. IF currentRatio IS High AND enterpriseToEbitda IS Very low AND priceToBook IS High THEN invest
2. IF currentRatio IS Low AND enterpriseToEbitda IS High AND priceToBook IS Low THEN invest
3. IF currentRatio IS Low AND enterpriseToEbitda IS Low AND priceToBook IS Medium THEN invest
4. IF currentRatio IS High AND enterpriseToEbitda IS High AND priceToBook IS High THEN invest
5. IF currentRatio IS Very high AND enterpriseToEbitda IS Very low AND priceToBook IS High THEN invest.

The full list of one hundred rules is presented in Appendix B.

Semantic representations of the linguistic values (Low, High, etc.) can be explained based on Figs. 4.16, 4.20, and 4.24.

The fuzzy IF-THEN rules for the full version of the recommendation problem, characterized by all the attributes presented in Sect. 4.3.2 and Appendix A, are included in Appendix C. Membership functions of fuzzy sets that can serve for semantic explanation of the recommender performance, along with these rules, are illustrated in Appendix E.

Fuzzy rules generated based on fuzzy data points are interpretable and explainable because they are directly based on each past transaction. One example creates one rule. Each rule differs in regard to the index of each past transaction. Therefore, the first five rules generated by the system reflect the pattern for all other rules derived from past transactions:

1. IF currentRatio IS currentRatio1 AND enterpriseToEbitda IS enterpriseToEbitda1 AND priceToBook IS priceToBook1 THEN invest
2. IF currentRatio IS currentRatio2 AND enterpriseToEbitda IS enterpriseToEbitda2 AND priceToBook IS priceToBook2 THEN invest
3. IF currentRatio IS currentRatio3 AND enterpriseToEbitda IS enterpriseToEbitda3 AND priceToBook IS priceToBook3 THEN invest
4. IF currentRatio IS currentRatio4 AND enterpriseToEbitda IS enterpriseToEbitda4 AND priceToBook IS priceToBook4 THEN invest
5. IF currentRatio IS currentRatio5 AND enterpriseToEbitda IS enterpriseToEbitda5 AND priceToBook IS priceToBook5 THEN invest.

It is worth emphasizing that the fuzzy clusters associated with the IF-THEN rules have been determined based on all data items in the dataset (concerning every user) while the fuzzy data points reflect only past transactions of a single user (investor). This is very important with regard to the explanation of the recommender performance.

Let us notice that all the rules (also those presented in Appendix C) have the same THEN part—with the conclusion "invest". As mentioned earlier, this is because the recommender is a one-class classifier. Such a system works based on the dataset from only one class. In contrary to a binary classifier that can produce two output decisions: recommend or not recommend, the one-class recommender can infer only one decision: recommend (invest).

It should be noted that the one-class classification is a classical ML (machine learning) problem that is recently often considered in the literature; see e.g. [13, 15, 18]. The one-class classifiers recognize instances of a concept by only using examples of the same concept. Instances of only a single object class are available during training. All other classes except the class given for training are called alien classes (or novel, abnormal, outlier classes). During testing, the classifier may encounter objects from alien classes. The goal of the classifier is to distinguish the objects of the known class from the objects of alien classes. A one-class classifier, where only examples of the positive class are available, applied in a recommendation system, within the framework of the collaborative filtering, is presented in [14]. The well-known ML methods, e.g., SVM (Support Vector Machines), k-means, and k-Nearest Neighbor have been modified to learn on the data from only one class (see e.g. [1]).

4.6 Results of the System Performance

This section concerns the performance of the recommender. Section 4.6.1 shows how recommendations are produced. Visualization of the results is presented in Sect. 4.6.2. Explanations of the recommendations are included in Sect. 4.6.3. Evaluation of the recommender's performance is illustrated in Sect. 4.6.4.

4.6.1 Recommendations Produced by the Recommender

The one-class recommender, proposed in this chapter, produces recommendations based on the data of past transactions of users (investors). As explained in Sect. 4.4, the data points that represent the past investments are viewed as fuzzy points in the attribute space. The fuzzy points are fuzzy sets defined by the Gaussian membership functions, as illustrated in Figs. 4.17, 4.21, 4.25, and Appendix F. An example of five rules formulated by use of the fuzzy sets shown in Figs. 4.17, 4.21, 4.25, and associated with transactions indexed as 1, 2, 3, 4, 5, are presented in Sect. 4.5. The rules of this kind are employed in order to recommend new investments; in two cases: for 3 and 21 attributes.

Let us call the rules created by the individual fuzzy points (past transactions)—recommendation rules, and the rules formulated by use of the fuzzy sets depicted in Figs. 4.16, 4.20, 4.24, and the rules included in Appendix C, with the fuzzy sets portrayed in Appendix E—explanation rules. The former rules are employed in order to produce recommendations while the latter is applied for explaining the system's decisions.

For particular users (investors), $u = 1, 2 \ldots, H$, and his past transactions, $t = 1, 2, \ldots, M_u$, let us present the recommendation rules, $R_t^{(u)}$, in the following, general form:

$$\textbf{IF } x_1 \text{ is } G_1^{(u,t)} \text{AND } x_2 \text{ is } G_2^{(u,t)} \text{ AND} \ldots \; x_n \text{ is } G_n^{(u,t)} \textbf{ THEN } \textit{Recommend} \quad (4.4)$$

where $G_i^{(u,t)}$ are Gaussian membership functions, $i = 1, 2, \ldots, n$, $u = 1, 2 \ldots, H$, $t = 1, 2, \ldots, M_u$, defined for every attribute of the past transactions for each user; n denotes the number of attributes, H—the number of users, M_u—the number of transactions of user u. It is obvious that $M_1, M_2, \ldots, M_H \leq M$ that is the number of data items in the dataset, composed of the past transactions of particular users $u = 1, 2, \ldots, H$; of course more than one user could realize the same transactions.

The explaination rules, R^k, for $k = 1, 2, \ldots, N$, are formulated as follows:

$$\textbf{IF } x_1 \text{ is } A_1^k \text{ AND } x_2 \text{ is } A_2^k \text{ AND} \ldots \text{ AND } x_n \text{ is } A_n^k \textbf{ THEN } \textit{Recommend} \quad (4.5)$$

where N denotes a number of the rules.

For each user, his past transactions are analyzed in order to produce new recommendations by the recommender. Every new data item that occurs is recommended to the user if there is a rule (one or more) activated by this data item with a sufficient degree. Therefore, the rule activation level should be determined.

Usually, the rule activation level, also called the rule firing level (see Sect. 2.3), is calculated by use of a T-norm operation, e.g. (2.8), as well as (3.17); most often the "product" or "min" T-norm is applied. However, with regard to the recommender considered in this chapter, and the problem described in Sect. 4.2, a different way of determining the rule firing level is proposed. Instead of the T-norm, the arithmetic average is employed; also, a weighted arithmetic average is suitable for this recommender.

When rules of the form (2.1) or (2.14) are used, with the "AND" operations in their antecedent part, the T-norm functions work very well to determine the rule firing level, in most cases of fuzzy systems applications. With regard to some problems, fuzzy IF-THEN rules can be formulated in a different way, e.g. with "OR" operators in the antecedent part. In such cases, the T-norm is not appropriate. The one-class recommender, considered in the application as an investment adviser, needs another way of aggregating particular terms in the antecedent part of the fuzzy IF-THEN rules (4.4), (4.5), in spite of the fact that the "AND" operation is employed. This is because the similarities of new recommendations to the past transactions do not necessarily require exact similarity with regard to every attribute. Therefore, the T-norm is proposed.

For example, in the case when a new recommendation is very similar to a past transaction (or almost the same) with regard to 22 attributes and differs much only on one attribute (assuming 23 attributes considered), the T-norm operator results in very low (or almost zero) value of the rule firing level. In such a situation, this new candidate for the recommendation will be rejected, and not recommended, despite the similarity with regard to the rest attributes. Therefore, the arithmetic average is more adequate. Hence, the following formula is applied in order to determine the rule firing levels:

$$\tau_t^{(u)} = \frac{1}{n}\sum_{i=1}^{n} G_i^{(u,t)}(\overline{x}_i) \tag{4.6}$$

for rules $R_t^{(u)}$; $u = 1, 2 \ldots, H, t = 1, 2, \ldots, M_u$, formulated as (4.4).

In the same way, the rule firing levels for rules R^k, $k = 1, 2, \ldots, N$, expressed as (4.5), are calculated as follows:

$$\tau_k = \frac{1}{n}\sum_{i=1}^{n} A_i^k(\overline{x}_i) \tag{4.7}$$

The main idea of the performance of the proposed one-class recommender is the application of the rule firing levels (4.6) as a measure of similarity of the new candidate for a recommendation to the past transactions of the user for whom the recommendation is generated.

4.6.2 *Visualization of the Recommender Results*

In Sect. 4.3.3, the t-SNE method for data visualization has been presented and applied. This method preserves the neighborhood of the data points, but distances between clusters are not preserved. Multidimensional data points are visualized in the 2D or 3D spaces. The axes of these spaces do not correspond to the attributes.

From this point of view, a similar idea is behind the novel method proposed by the author of this book, with regard to visualization of the recommendations generated by the one-class recommender. As a matter of fact, not only the visualization but the algorithm of inferring the recommendations is novel. It should be emphasized that the recommender that performs by use of this new method is totally interpretable and explainable.

As concluded at the end of Sect. 4.6.1, in the method proposed in this chapter, the recommender offers a new recommendation for a user, if it is similar to his past transactions. The similarity is measured by the use of the rule firing levels, expressed by Eq. (4.6).

Let us consider a new data point that is a candidate for a recommendation for a user (investor). Figure 4.26 portrays this point as a yellow dot at the center. All other data points that represent past transactions of this user are illustrated in this figure and correspond to their firing levels by the quantity

$$1 - \tau_t^{(u)}$$

Every past transaction, as well as the new candidate for the recommendation, are multidimensional points characterized by the attributes described in Sect. 4.3.2 and Appendix A. The visualization of these points in the 2D space, in Fig. 4.26, shows only how far nearest neighbors are located from the yellow point. However, the real distances are not depicted. In other words, this visualization reflects similarities to past transactions.

As mentioned earlier, the similarities are measured by the use of the rule firing levels. Every point at this figure, representing a past transaction of the user, is viewed as a fuzzy point—fuzzy numbers with Gaussian membership functions in particular attribute domains, as described in Sect. 4.4. Each of the past transactions refers to one fuzzy IF-THEN rule of the form (4.4); examples of such rules are formulated in Sect. 4.5. Firing levels of these rules, calculated by Eq. (4.6), where $\overline{x}_i$, for $i = 1, 2, \ldots, n$, corresponds to the new candidate for the recommendation (the yellow point), determine the similarities.

For a better visualization, an adjustment function, defined as presented in Fig. 4.27 has been proposed and applied, resulting in the corrected (adjusted) visualization of the same points. Hence, the points portrayed in Fig. 4.26 are presented as Fig. 4.28 illustrates. Of course, this transformation has been introduced only for better visualization. The distances have not been changed. The points located beyond those gathered at the circle are far away, so definitely not considered as neighbors. Figures 4.26 and 4.28 show the visualizations of the most similar past transactions realized by IBM as an investor to recommendation no. 187.

Other examples of the visualizations are presented in Figs. 4.29 and 4.30 for recommendation no. 363 and investor with ticker GOOG, as well as in Figs. 4.31 and 4.32 for recommendation no. 1545 and investor with ticker MOTS; not adjusted and adjusted versions, respectively.

Based on the visual representation of firing levels of the rules corresponding to past transactions, it is very intuitive to compare candidates and infer which recommendation is better, as presented in Fig. 4.33.

As we see, the candidate for the recommendation no. 363 has one close neighbor (similar point—denoted as green). However, it seems that the distance between the yellow and green points is smaller in Fig. 4.29 than in Fig. 4.30, it is not true; the distance is the same from the yellow point to the green one in both figures, and this also concern distances from the yellow point to others. Thus, the similarities are the same but visualized in a different way. Analogously, in Figs. 4.31 and 4.32, both visualizations say that recommendation no. 1545 has many more neighbors (similar points).

Table 4.5 presents values of firing levels of rules associated with the nearest points of past transactions for the recommendation no. 1545; see Figs. 4.31 and 4.32. For recommendation no. 363, in Figs. 4.29 and 4.30, the rule firing level of the nearest neighbor (the green point) is much higher, equal 0.9037494717563561. The rule associated with this point (no. 364) has the following form:

IF currentRatio IS currentRatio364 AND earningsPerShareGrowth IS earnings PerShareGrowth364 AND revenueGrowth IS revenueGrowth364 AND currentAssets IS currentAssets364 AND profitMargins IS profitMargins364 AND debtToE-

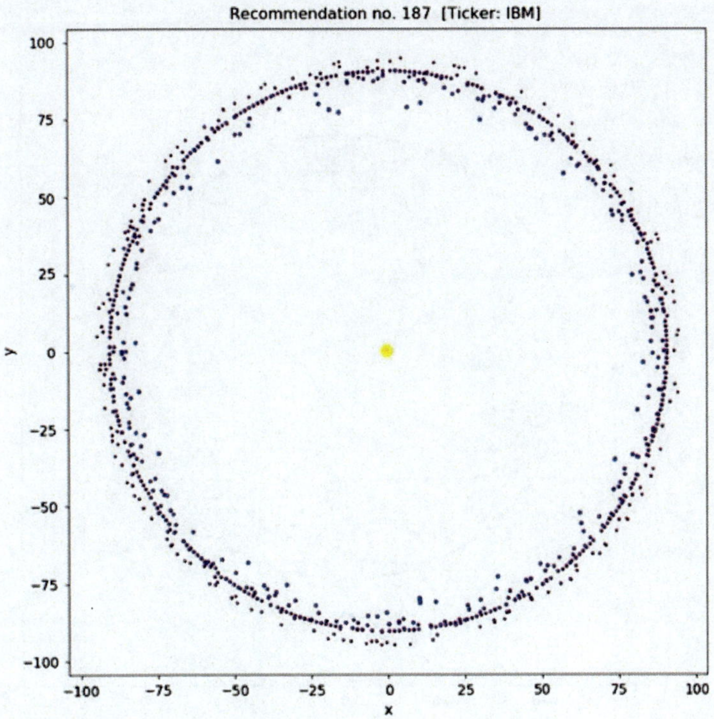

Fig. 4.26 Recommendation 187—visualization of the surrounding neighbors before adjustment

x	y
0.5	0.95
0.4	0.9
0.3	0.85
0.2	0.7
0.1	0.6
0.05	0.4
0.01	0.001

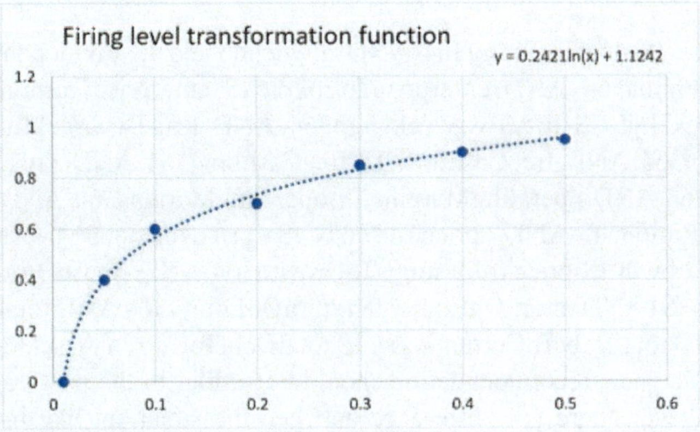

Fig. 4.27 Adjustment function

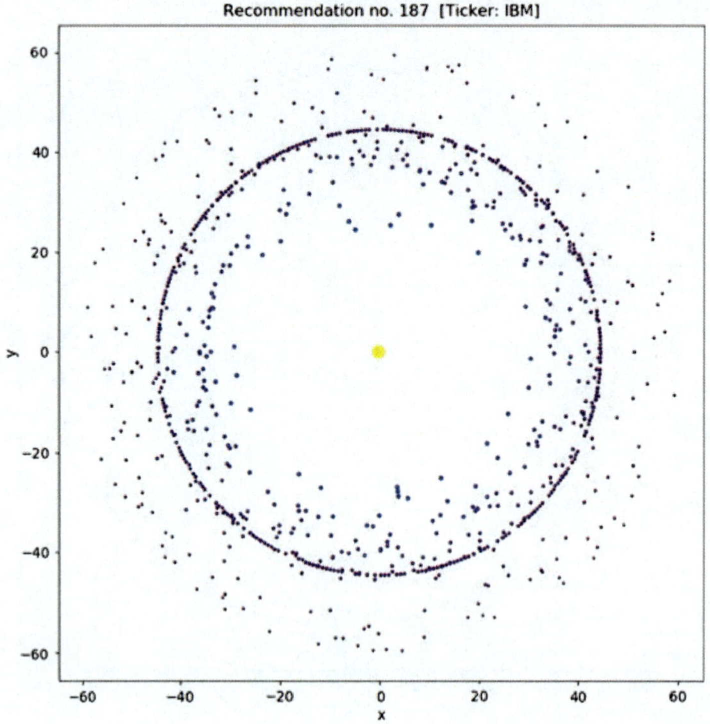

Fig. 4.28 Recommendation 187—visualization of the surrounding neighbors after adjustment

quity IS debtToEquity364 AND dividendYield IS dividendYield364 AND ebitda IS ebitda364 AND enterpriseToEbitda IS enterpriseToEbitda364 AND fiftyTwo WeekHigh IS fiftyTwoWeekHigh364 AND fiftyTwoWeekLow IS fiftyTwoWeek-Low364 AND freeCashflow IS freeCashflow364 AND grossProfits IS grossProfits364 AND operatingMargins IS operatingMargins364 AND payoutRatio IS payoutRatio364 AND priceToBook IS priceToBook364 AND priceToEarnings ToGrowth IS priceToEarningsToGrowth364 AND returnOnAssets IS returnOnAssets364 AND returnOnEquity IS returnOnEquity364 AND totalDebt IS totalDebt364 AND totalDebtToCurrentAsset IS totalDebtToCurrentAsset364 AND THEN invest.

The new recommendation should be similar to the past transactions of the user. However, there is a difference between the situation, like in Figs. 4.29 and 4.30, where only one neighbor with very high firing level exists, and the case illustrated in Figs. 4.31 and 4.32. With regard to the candidate for recommendation no. 1545 there is a group of neighbors (no. 1716, no 885, …, no. 1992), with firing levels included in Table 4.5. Although the data point no. 1545 is much more similar to the yellow point (in Figs. 4.31 and 4.32), this is the only one such a neighbor. From the recommendation point of view, it is important to take into account a group of neighbors (past transactions) even less similar to the yellow point

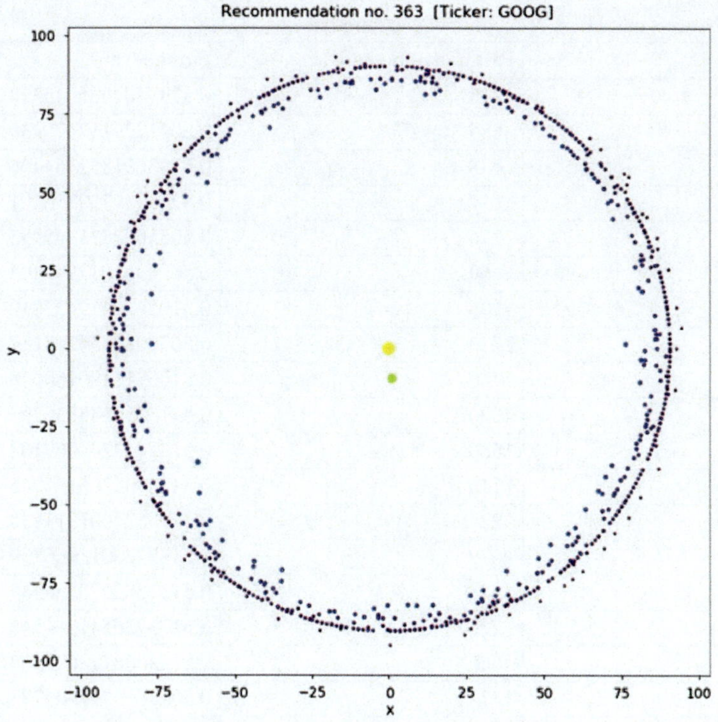

Fig. 4.29 Recommendation 363—visualization of the surrounding neighbors before adjustment

(like in Figs. 4.29 and 4.30). Therefore, the alignmentWithStrategy function is now proposed, expressed as follows:

$$alignmentWithStrategy = \mathbf{max}(allFiringLevelsAboveAlignmentThreshold) + numberOfNeighbors/100 - 0.01 \quad (4.8)$$

where $allFiringLevelsAboveAlignmentThreshold$ is a number of all data points (past transactions of a single user) with values of the firing levels above the threshold that is set for this algorithm. The $numberOfNeighbors$ denotes the number of neighbors; for example 1 in the case illustrated in Figs. 4.29 and 4.30, and 35 for the group of neighbors presented in Figs. 4.31 and 4.32, as well as in Table 4.5.

The $alignmentWithStrategy$ function (4.8) defines the similarity of a candidate point for the recommendation to the past transaction points of a considered user (investor), taking into account a number of the similar points (neighbors).

The Python code, which has been used to locate the points in the visualizations, illustrated in the form of the above-considered figures, is presented as Algorithm 1. It shows how the points are visualized within the circle shape, depending on the threshold with regard to the firing levels.

Table 4.5 Nearest neighbors of recommended item (stock 1545)

	Past transaction no	Firing level
1	1716	0.5237704986545398
2	885	0.5061329117676939
3	448	0.5343601852464406
4	17	0.5233999739756401
5	276	0.5055007871445953
6	69	0.5078961017999403
7	1105	0.5105920419422019
8	1723	0.5030988354218137
9	1940	0.5105911190160048
10	2000	0.541375014883629
11	1580	0.5142442247069102
12	1743	0.5435931776197946
13	224	0.5773524001313838
14	679	0.5189220018966799
15	343	0.5137683970556439
16	275	0.5173726541214146
17	666	0.5164666728675236
18	1272	0.5017656997705773
19	1068	0.5189467725780076
20	1687	0.54506645839053
21	996	0.638253720681402
22	138	0.5381266702500033
23	1825	0.5874929881456701
24	982	0.5876084606172998
25	1356	0.5411139053492937
26	1941	0.5215412566984599
27	1937	0.5062740171859651
28	810	0.5173135457033435
29	110	0.5381266702500033
30	1183	0.5091763045498241
31	60	0.5232612366313513
32	150	0.5468779210439163
33	1401	0.519227787049489
34	461	0.5301363070687621
35	1992	0.5551927641240275

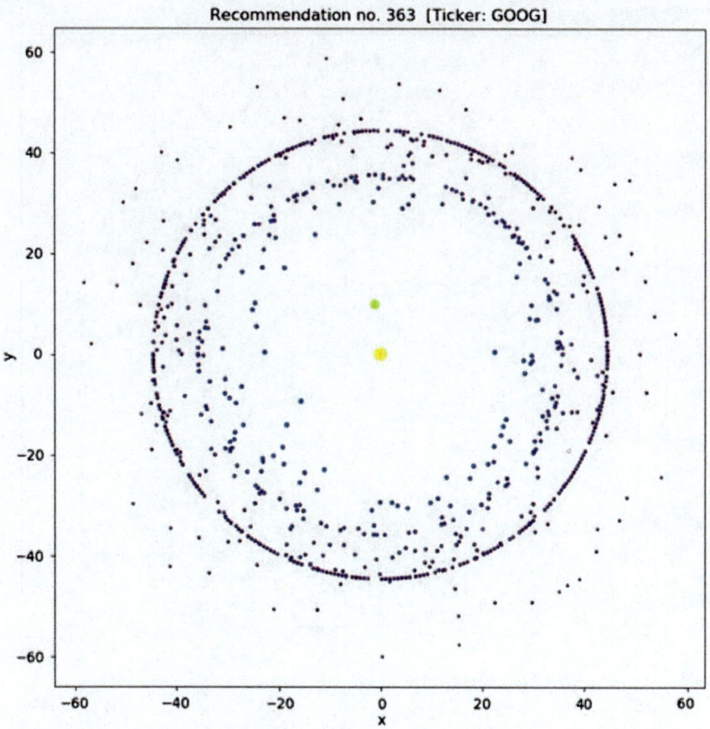

Fig. 4.30 Recommendation 363—visualization of the surrounding neighbors after adjustment

Algorithm 1 Python code

```
1: for i, firingLevel in enumerate(allFiringLevels):
2: adjustedFiringLevel = 0.2421 * math.log(firingLevel, math.e) + 1.1242
3: x = 100*(1-firingLevel)*math.cos(int((360/len(allFiringLevels))*i))
4: y = 100*(1-firingLevel)*math.sin(int((360/len(allFiringLevels))*i))
```

4.6.3 Explanations of the Recommendations

As presented in Sect. 4.5 and formally described in Sect. 4.6.1, explanation rules (4.5) are applied in order to explain why the recommender decides to offer the recommendations. The fuzzy IF-THEN rule activated with the highest level is employed to provide the explanation with regard to the semantic meaning of the fuzzy sets in this rule. This refers to the linguistic labels of the fuzzy sets, such as "Low", "High", "Very high", etc. The explanation rules are considered along with the visualizations proposed in Sect. 4.6.2, taking into account the $alignmentWithStrategy$ function (4.8).

Let us consider a data point with the values of the membership functions for particular attributes as presented in Table 4.6. In this case: firing level =

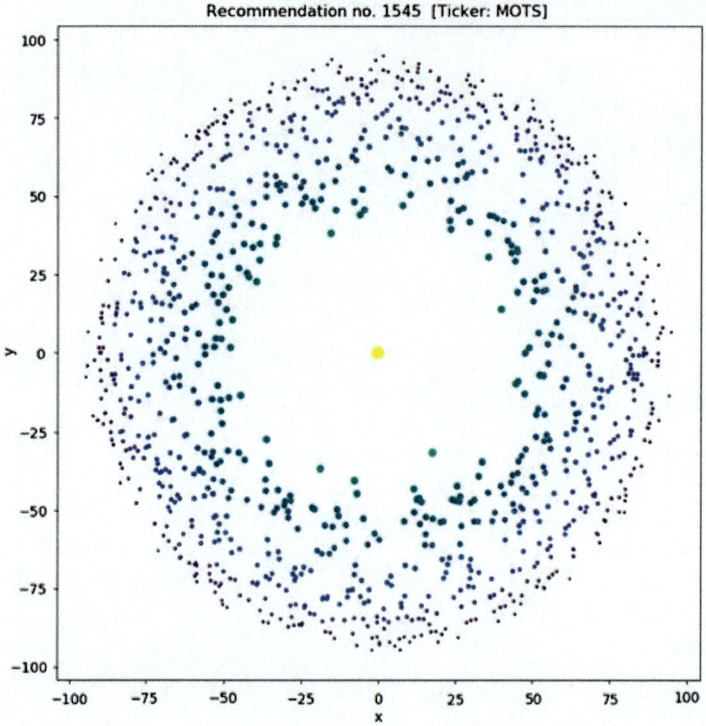

Fig. 4.31 Recommendation 1545—visualization of the surrounding neighbors before adjustment

0.638253720681402, number of neighbors = 37, and the value of fuction (4.8) = 0.998253720681402.

The following fuzzy IF-THEN rule is applied in order to explain the recommendation of the data point described in Table 4.6.

IF currentRatio IS Very high AND earningsPerShareGrowth IS Low AND revenueGrowth IS Very high AND currentAssets IS Very low AND profitMargins IS Very low AND debtToEquity IS Very low AND dividendYield IS Very low AND ebitda IS Low AND enterpriseToEbitda IS Very low AND fiftyTwoWeekHigh IS Low AND fiftyTwoWeekLow IS Very low AND freeCashflow IS Low AND grossProfits IS Very low AND operatingMargins IS Very low AND payoutRatio IS Very low AND priceToBook IS High AND priceToEarningsToGrowth IS Very low AND returnOnAssets IS Very low AND returnOnEquity IS Very low AND totalDebt IS Very low AND totalDebtToCurrentAsset IS Very low AND THEN invest.

This rule has been activated with the values of the membership functions, and the firing level, included in Table 4.6. Let us notice that the value of the $alignmentWithStrategy$ function is very high because of many neighbors (37 similar data points of the past transactions).

Table 4.7 includes values of membership functions of fuzzy sets in the IF-THEN rules, for 3 input attribute values, that characterize selected companies which names

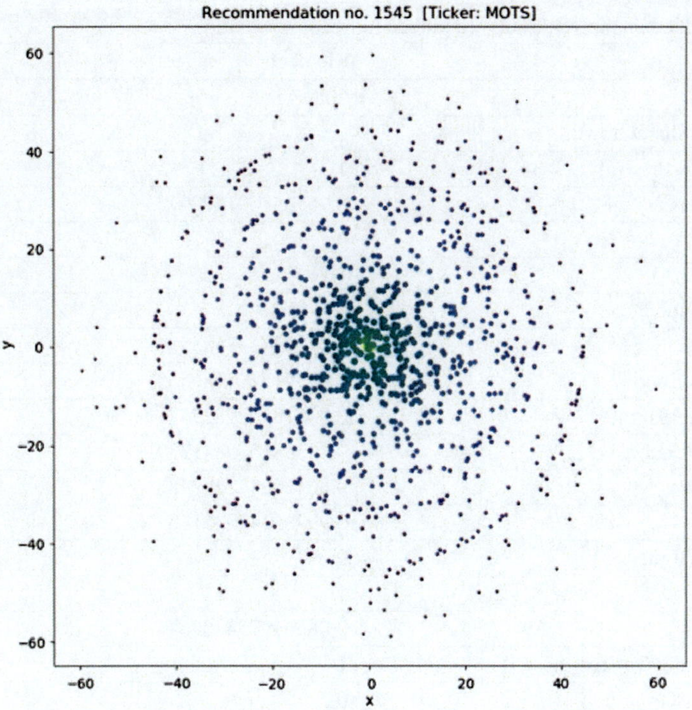

Fig. 4.32 Recommendation 1545—visualization of the surrounding neighbors after adjustment

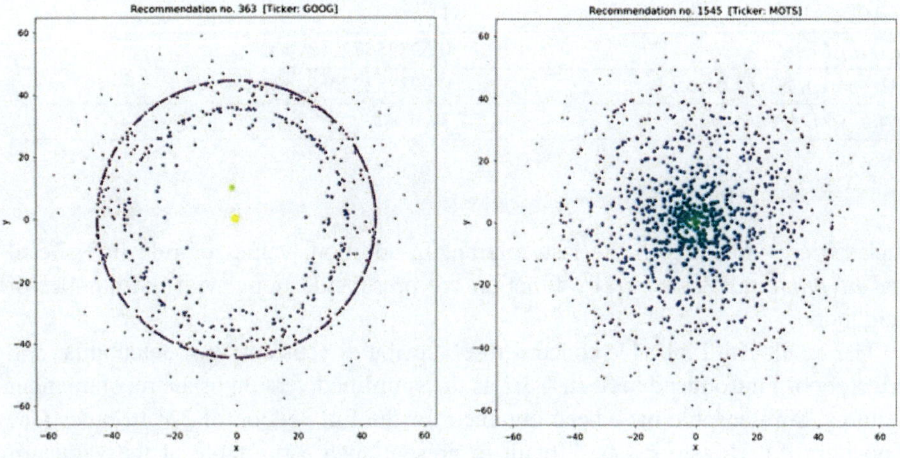

Fig. 4.33 Comparing recommendation no. 363 and no. 1545

Table 4.6 Membership values of a candidate for recommendation

Attribute	Membership value
Currentratio	208
EarningsPerShareGrowth	0
RevenueGrowth	1
CurrentAssets	0.965638091279073
ProfitMargins	0
DebtToEquity	1
DividendYield	1
Ebitda	0.942934143585586
EnterpriseToEbitda	0.455715294986565
FiftyTwoWeekHigh0	0.90518199982567
FiftyTwoWeekLow	0.243881615841239
FreeCashflow	0.804608441492486
GrossProfits	0.99985447996148
OperatingMargins	0
PayoutRatio	1
PriceToBook	0.086699754409646
PriceToEarningsToGrowth	1
ReturnOnAssets	0
ReturnOnEquity	0
TotalDebt	0.998814312927695
TotalDebtToCurrentAsset	1
$FiringLevel$	0.638253720681402
$AlignmentWithStrategy$	0.998253720681402
$NumberOfNeighbors$	37

(tickers) are indicated in the last column. In addition, values of rule firing level and $alignmentWithStrategy$ function are presented, along with the number of neighbors.

The results in Table 4.7 concern three attributes (currentratio, evtoebitda, and pricetobook) introduced in Sect. 4.3.2 as the simplified version of the recommender features. Similar table have been obtained for the full version of 21 attributes (see Appendix A). However it is difficult to present such a big table of the values of membership functions, firing levels, $alignmentWithStrategy$ function, and number of neighbors for the recommender with 21 inputs. Therefore, only fragments are illustrated in Table 4.8.

Based on the calculated alignment, degrees of membership for each feature and the distance between the closest neighbors, it is possible to generate explanations that are easy to understand by a user:

Table 4.7 Rule firing levels and values of membership functions, for 3 attributes

	currentRatio	enterpriseToEbitda	priceToBook	firingLevel	alignmentWithStrategy	numberOfNeighbors	rowIndex	ticker
171	1.00000	1.00000	1.00000	1.00000	2.99000	200	1627	ZION
99	1.00000	1.00000	1.00000	1.00000	1.76000	77	1677	EQT
53	1.00000	1.00000	1.00000	1.00000	3.17000	218	1563	PACW
212	1.00000	1.00000	1.00000	1.00000	1.90000	91	2075	LSXMA
156	1.00000	1.00000	1.00000	1.00000	3.11000	212	2105	PNC
354	1.00000	1.00000	1.00000	1.00000	1.33000	34	23	SPEX
20	1.00000	1.00000	0.99761	0.99920	3.01920	203	394	JPM
357	1.00000	1.00000	0.99761	0.99920	3.02920	204	2125	STT
16	1.00000	1.00000	0.99761	0.99920	3.15920	217	1459	DFS
275	1.00000	1.00000	0.99049	0.99683	3.06683	208	1427	BK
323	1.00000	1.00000	0.99049	0.99683	3.04683	206	937	HBAN
149	1.00000	1.00000	0.99049	0.99683	3.15683	217	318	CI
213	1.00000	1.00000	0.99049	0.99683	1.41683	43	775	LBRDA
169	1.00000	0.99353	0.97874	0.99075	3.36075	238	822	STI
2	1.00000	0.97918	0.99049	0.98989	2.11989	114	539	AL
22	1.00000	1.00000	0.96251	0.98750	2.97750	200	369	GS
396	0.94993	0.99974	1.00000	0.98322	2.64322	167	1369	TPX
142	1.00000	1.00000	0.94205	0.98068	3.09068	212	1471	ETFC
85	1.00000	0.93566	0.99761	0.97776	3.21776	225	1042	ACGL
376	0.94993	0.96905	1.00000	0.97300	2.45300	149	1439	CL
283	1.00000	0.89175	0.99049	0.96075	3.23075	228	1324	PACW
158	1.00000	0.98735	0.88957	0.95898	3.18898	224	743	ESQ
157	1.00000	0.87159	0.99761	0.95640	2.89640	195	1432	BRK.B
340	1.00000	0.99974	0.85827	0.95267	2.95267	201	1202	BBT
405	0.94993	0.90130	0.96251	0.93791	2.10791	118	1763	AAP
297	0.94993	0.86102	1.00000	0.93698	2.61698	169	985	PZZA
276	0.89086	0.92767	0.99049	0.93634	2.06634	114	1675	EPE
162	1.00000	0.82746	0.97874	0.93540	3.24540	232	1357	STI
373	1.00000	0.79151	0.99049	0.92733	3.13733	222	164	ETFC
24	0.94993	0.96905	0.85827	0.92575	2.36575	145	1288	LAD
370	0.81428	0.99766	0.96251	0.92482	1.67482	76	146	CRL
84	0.81428	0.99585	0.91762	0.90925	2.05925	116	1726	CLR
385	1.00000	0.71412	0.99761	0.90391	3.04391	215	62	ANTM
128	1.00000	1.00000	0.70902	0.90301	2.84301	195	457	SCHW
58	1.00000	0.77906	0.91762	0.89890	3.07890	219	1211	C
148	0.72540	0.99069	0.97874	0.89828	2.06828	118	1123	PE
250	0.98724	0.71412	0.96251	0.88796	2.14796	127	692	ADNT
211	1.00000	0.65989	1.00000	0.88663	2.78663	191	2085	MTG
82	1.00000	0.71412	0.94205	0.88539	3.08539	221	1647	RNR
368	1.00000	0.64615	1.00000	0.88205	2.94205	207	1312	MTG
28	0.98724	0.75358	0.88957	0.87680	2.26680	140	348	FBHS
143	0.81428	0.97918	0.82413	0.87253	1.90253	104	338	DVN
236	0.72540	0.99974	0.88957	0.87157	2.43157	157	955	KNX
48	1.00000	0.82746	0.74905	0.85884	2.82884	198	595	AFL
111	0.98724	0.83893	0.74905	0.85841	2.34841	150	1291	LH
34	0.72540	0.92767	0.91762	0.85690	2.05690	121	1440	CLS
152	1.00000	0.56333	1.00000	0.85444	3.15444	231	1330	PNC
19	1.00000	0.59091	0.96251	0.85114	3.11114	227	1629	ALL
219	0.72540	0.82746	0.99049	0.84779	1.87779	104	441	QUAD
117	0.72540	0.85012	0.96251	0.84601	1.71601	88	1103	CRZO
404	0.81428	0.75358	0.96251	0.84346	1.78346	95	1015	TRCO
402	0.98724	1.00000	0.54263	0.84329	2.14329	131	387	IR
181	1.00000	0.85012	0.66793	0.83935	1.96935	114	572	PDCE
327	0.53301	0.95704	1.00000	0.83002	2.11002	129	1368	TOT
153	0.94993	0.57710	0.96251	0.82985	1.69985	88	550	CLR

This stock might be interesting for you since operating margin is very high and return on assets is medium. Moreover, return on equity is extremely high, and current ratio is low.

Here are your most relevant past transactions that helped surface this recommendation: LVS in Dec17, SYY in Mar17, LVS in Mar18.

Table 4.8 Rule firing levels and values of membership functions, for 21 attributes

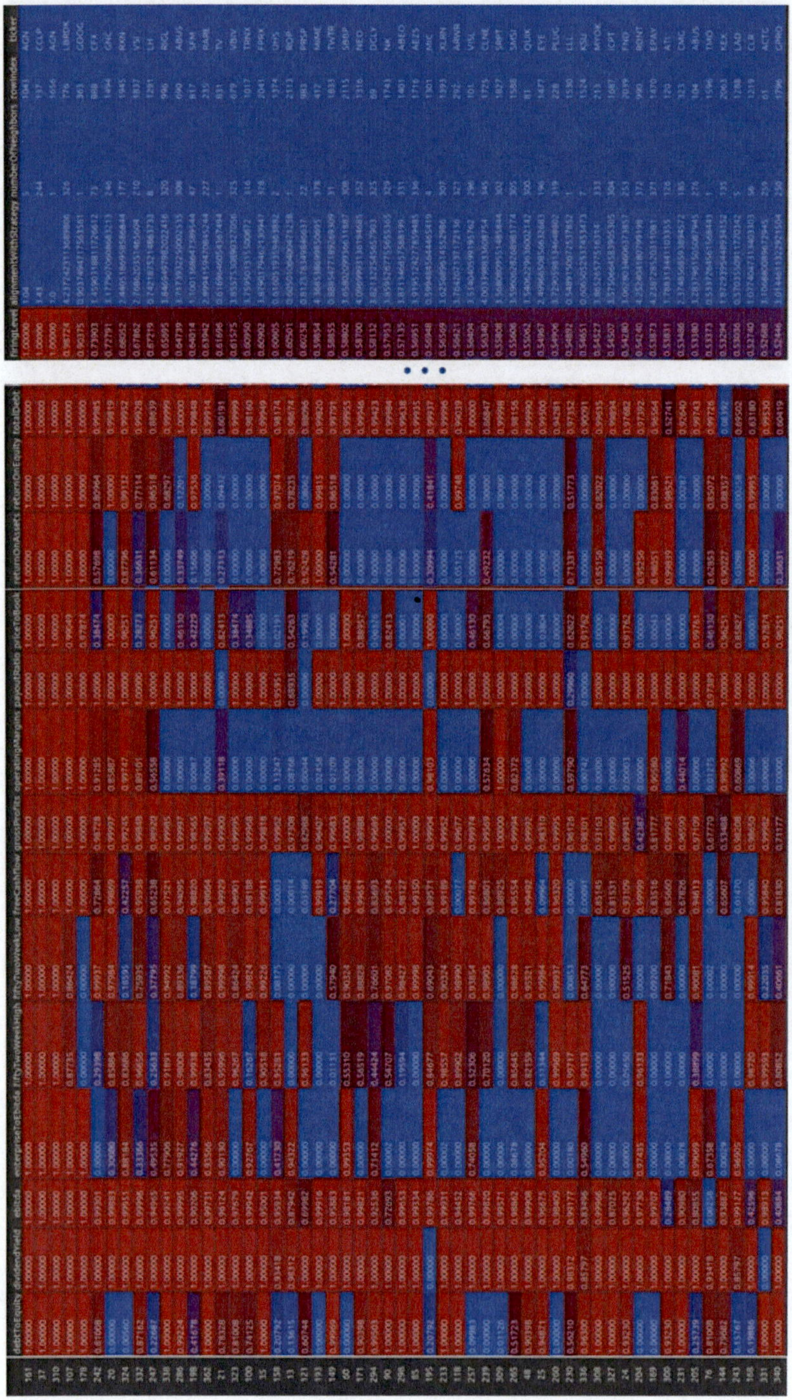

4.6.4 *Evaluation of the Recommender Performance*

In Chap. 3, the RMSE (Root Mean Squared Error) has been used in order to evaluate the performane of the recommenders. The RMSE is a very popular measure for evaluating the accuracy of predicted ratings. This is usually applied in ML (Machine Learning), including multi-classification problems, with labeled data.

With regard to the one-class recommender, considered in this chapter, the RMSE cannot be applied. In this case, the "Recall"—also known as "Sensitivity" or "True Positive Rate"—is employed. This measure is based on the values of the TP—"True Positive" and FN—"False Negative", well known in the literature, along with FP—"False Positive" and TN—"True Negative"; see e.g. [16]. The "Recall" measures the proportion of predicted (recommended) positive cases to really existed positive cases (investments).

The "Recall", as the measure of performance of the recommender, is included in Tables 4.10 and 4.11. The history of transactions realized by different funds (banks and other investors), with results of recommendations produced by the one-class recommender, is presented. These tables include selected examples of a large number of funds considered.

The meaning of particular columns is explained in Table 4.9. The number of all stocks from a validation quarter equals 4476; see column "All". The first column (All R) indicates the number of all stocks recommended in the validation quarter by the recommender. The second column (All B) shows the number of all stocks bought in the validation quarter by particular funds. Next columns denotes: B&R—stocks that were bought and recommended, B~R—stocks that were bought but not recommended. The former equals the TP (True Positive); the latter equals FN (False Negative). Let us notice that we have only positive data available for the one-class recommender. Thus, the number of R~B—stocks that were recommended but not bought (FP—False Positive), cannot be used for the performance evaluation. It is possible that the system offers good recommendations, but we do not know why the recommended stocks have not been bought (no data from the negative class). For the

Table 4.9 Columns in Tables 4.10 and 4.11

Fund name	Name of funds
All R	All stocks Recommended in a validation quarter
All B	All stocks Bought in a validation quarter
All	All stocks from a validation quarter
B&R	Stocks that were Bought and Recommended (TP)
B~R	Stocks that were Bought but Not Recommended (FN)
Recall	TP/(TP+FN)

same reasons, the TN (True Negative) cannot be taken into account. Therefore, only the "Recall" is used as a measure of the recommender's performance.

The last column in Tables 4.10 and 4.11 includes values of the "Recall", calculated as TP/(TP+FN) where TP and FN denote "True Positive" and "False Negative", respectively; see e.g. [16].

Tables 4.10 and 4.11 illustrate the "Recall" values for selected Funds. It should be admitted that usually, this value is not higher than included in these tables, or even lower. However, it is not bad if we realize that the fact that the recommender offers a recommendation that has not been used (stocks not bought) does not mean that this recommendation is not adequate. In many cases, the recommender produces recommendations realized by investors as past transactions.

It should be emphasized that having only positive data points (all stocks bought by investors), the fact that the recommender proposes a recommendation for a particular investor, who is not interested (does not invest), is nothing wrong. This recommendation may be good for this user since another investor has realized this recommendation.

The recommender produces recommendations based on past transactions of an investor. Therefore, it is also understandable that the system does not infer the decision to "invest" with regard to the data point that is not similar to the previous investments of this user.

4.7 Conclusions Concerning the Proposed One-Class Recommender

The main difference between the recommenders proposed in Chap. 3, with application to the MovieLens data, and this one—created to support investment advisers—concerns the performance of the systems as classifiers. With regard to the movie recommendations, the neuro-fuzzy systems (as recommenders A, B, or C) play a role of a multi-class classifier—when data with labels that denote ranking of the movies, for example, 2, 3, 4, 5, are used. Thus, we have the data labeled as belonging to different classes. A special case is a binary classification, when two classes are considered, e.g. "Yes" and "No". The recommender presented in this chapter is viewed as a one-class classifier. This means that all the learning (training) data, employed in order to generate fuzzy IF-THEN rules, are assigned to one class. Hence, every rule has the same consequent part (THEN invest).

Another difference concerns the data. Apart from the fact that the MovieLens dataset contains nominal (categorical) data that should be encoded into numerical values, the preference data (ratings) comes from different sources. According to [7], with regard to the preference data, two main cases are distinguished: explicit and implicit ratings. The former means preferences that the user has explicitly stated for particular items, e.g., the ratings in the MovieLens dataset, which is a user-provided star rating from 0.5 to 5 stars (or with a granularity of 1-star, in the older version of

Table 4.10 Results of recommendations for different funds; part 1

Fund name	All R	All B	All	B & R	B'R	Recall %
ADAGE CAPITAL PARTNERS GP LLC	265	226	4476	43	183	19,03
ADVISORY RESEARCH INC	467	96	4476	21	75	21,88
Baird Financial Group, Inc.	188	362	4476	89	273	24,59
BALDWIN BROTHERS INC MA	103	75	4476	19	56	25,33
Balentine LLC	190	152	4476	26	126	17,11
BANK OF HAWALL	121	89	4476	19	70	21,35
BANK OF NOVA SCOTIA	373	371	4476	148	223	39,89
Baystate Wealth Management LLc	83	144	4476	20	124	18,00
BB&T CORP	119	200	4476	36	164	18,00
BBVA COMPASS BANCSHARES, INC	161	78	4476	19	59	24,36
BEJAMIN F. EDWARDS & COMPANY, INC.	161	367	4476	57	310	15,53
BKD Wealth Advisors, LLc	112	106	4476	30	76	28,30
Bollard Group LLC	83	143	4476	19	124	13,29
BOSTON ADVISORS LLC	190	164	4476	31	133	18,90
BRINKER CAPITAL INC	114	187	4476	20	167	10,70
Brookstone Capital Management	135	71	4476	22	49	30,99
Brown Advisory Securities, LLC	114	86	4476	24	62	27,91
BTIM Corp.	280	80	4476	26	54	32,50
BURINEY CO	73	132	4476	16	116	12,12
C M BIDWELL & ASSOCIATES LTD	350	107	4476	33	74	30,84

(continued)

Table 4.10 (continued)

Fund name	All R	All B	All	B & R	$\bar{B}R$	Recall %
CALAMOS ADVISORS LLC	182	224	4476	62	162	2768
Campbell & CO Investment Adviser LLC	173	203	4476	21	182	10,34
CANANDAOGUA NATIONAL BANK & TRUST CO	235	108	4476	34	74	31,48
CAPITAL INTERNATIONAL SARL	143	86	4476	20	66	23,26
Capital Investment Advisors, LLC	501	184	4476	134	50	72,83
CAPROCK Group, Inc.	22	252	4476	90	162	35,71
CAPTRUST FINANCIAL ADVISORS	203	742	4476	126	616	16,98
Cetera Advisors LLC	262	140	4476	50	90	35,71
CHARTWELL INVESTMENT PARTNERS, INC.	579	157	4476	59	98	37,29
Checchi Capital Advisers, LLC	211	34	4476	12	22	35,29
Chevy Chase Trust Holding, Inc.	215	225	4476	85	140	37,78
Clearac Capital Inc	181	5	4476	3	2	60,00
COHEN & STEERS INC	234	111	4476	35	76	31,53
COLDSTREAM CAPITAL MANAGEMENT INC	134	100	4476	28	72	28.49
COLONY GROUP LLC	173	78	4476	23	55	29,49
COLUMBUS CIRCLE INVESTORS	402	60	4476	21	39	35,00
COMERICA SECURITIES INC	73	109	4476	17	92	15,60
COMMERCE BANK	111	276	4476	45	231	16,30

Table 4.11 Results of recommendations for different funds; part 2

Fund name	All R	All B	All	B & R	$B\bar{R}$	Recall %
Ledyard National Bank	55	97	4476	20	77	20,62
LEVIN CAPITAL STRATEGIES, L.P.	401	80	4476	20	60	25,00
Linscomb & Williams, Inc.	123	122	4476	26	96	21,31
Livforsakringsaktiebolaget Skandia (publ)	109	86	4476	14	72	16,28
Lloyds Banking Group plc	372	16	4476	4	12	25,00
LOGAN CAPITAL MANAGEMENT INC	74	124	4476	16	108	12,90
Lombard Odier Asset Management (Switzerland) SA	78	136	4476	20	116	14,71
MetLfe Investment Advisors, LLc	1209	138	4476	63	75	45,65
NEW MEXICO EDUCATIONAL RETIREMENT BOARD	675	24	4476	10	14	41,67
NISSAY ASSET MANAGEMENT CORP JAPAN ADV	592	373	4476	190	183	50,94
PARADIGM ASSET MANAGEMENT CO LLC	292	68	4476	32	36	47,06
PGGM Investments	320	45	4476	18	27	40,00
PINNACLE FINANCIAL PARTNERS INC	348	74	4476	46	28	62,16
Robert Olstein	387	94	4476	38	56	40,43
ROTHSCILD INVESTMENT CORP IL	378	68	4476	31	37	45,59
ROYAL LONDON ASSET MANAGEMENT LID	353	410	4476	172	238	41,95
Three Peaks Capital Manageement, LLc	420	72	4476	34	38	47,22
TWIN CAPITAL MANAGEMENT INC	286	40	4476	15	25	37,50
WHITTIER TRUST CO	413	276	4476	85	191	30,80
Winfield Associates, Inc.	159	7	4476	3	4	42,86
ZACKS INVESTMENT MANAGEMENT	254	135	4476	25	110	18,52
Zurich Insurance Group Ltd FI	73	138	4476	16	122	11,59

the data). The latter is inferred by the system from observable user activity, such as purchases of clicks. This (implicit rating) is realized in the recommender system for investment advisers, proposed in this chapter, where the preference data are inferred based on the history of transactions realized by users.

There is also a difference concerning the accuracy that is commonly used to evaluate the performance of recommendation systems. The RMSE (Root Mean Square Error) is applied for the recommenders presented in Chap. 3. This is the measure of accuracy, typically employed for the case of the explicit ratings (see Chap. 1). When ratings are not available, measuring the rating prediction accuracy is not possible. In such cases, the problem of finding the best item is usually transformed into the task of recommending to a user a list containing a certain number of items likely to interest him or her [4, 17]. Thus, the performance of the recommender described in this chapter is evaluated based on the "Recall" measure, included in Tables 4.10 and 4.11.

However, both kinds of the recommendation systems (presented in this chapter and Chap. 3) work within the content-based approach (see Chap. 1, Sect. 1.3). This means that the systems produce recommendations based on the attributes that characterize objects (items). As explained in Chap. 1, this paradigm differs from the collaborative filtering—that uses historical data (ratings) to determine user or item similarities. The recommenders in the content-based approach employ attribute values in order to determine users' preferences. Then, the systems offer recommendations according to the preferences of individual users (not based on similarities to other users).

As the one-class classifier, the recommender proposed in this chapter can be viewed as a special case of neuro-fuzzy recommenders, that is, the rule-based system—knowledge-driven. As a matter of fact, the neuro-fuzzy systems, considered in Chaps. 2 and 3, as combinations of fuzzy systems and neural networks, are knowledge-driven and also data-driven. The fuzzy IF-THEN rules are generated from data by the use of the WM method.

The difference, however, refers to the learning of the neuro-fuzzy systems. The recommenders presented in Chap. 3 have been trained, by use of the learning procedure that is similar to the backpropagation (commonly applied in multilayer feedforward neural networks). Thus, the membership functions of fuzzy sets in the rules are being tuned to achieve better performance of the system. Of course, there is a tradeoff between accuracy and interpretability (see Sect. 1.4). Therefore, with the recommenders presented in Chap. 3 the learning has been realized in the way to preserve the interpretability.

The explanation of the performance of each proposed recommender is very important and possibly based on the fuzzy IF-THEN rules, with fuzzy sets associated with linguistic labels having semantic meanings. In this way, the explainable recommenders, along with their recommendations, also produce appropriate explanations of the decisions.

The one-class recommender, presented in this chapter, had not been trained. The fuzzy IF-THEN rules have been formulated based on the data, analyzing histograms (see Figs. 4.14, 4.18, 4.22, and Appendix D). The additional tuning is not possible because of having the data belonging to only one class. In this case, we can say that

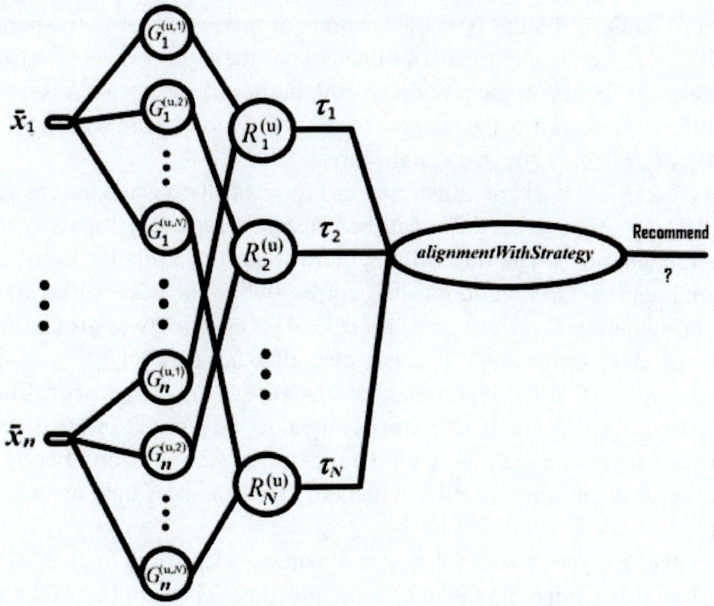

Fig. 4.34 Neuro-fuzzy recommender for a particular user

the knowledge represented by the fuzzy IF-THEN rules comes from both the data and also from knowledge from the experts. Therefore, it is not necessary to apply any learning algorithm.

As the special case of neuro-fuzzy systems, the one-class recommender, as a matter of fact, is confined to a fuzzy system. However, it can still be viewed as the neuro-fuzzy network of the form illustrated in Fig. 2.2, in Chap. 2. The number of inputs and outputs of this network corresponds to the numbers of attributes and rules, respectively. Thus, two architectures of the neuro-fuzzy recommender are considered: with 3 and 21 inputs, with regard to the recommendations for investment advisers, based on the real data from the asset management companies. In the case of the one-class classifier, the recommendation (or recommendations) is inferred from the output (outputs) with maximal value (values). This means that the decision "invest" comes from the rule (rules) activated with a maximum of the firing level value. The higher this value is, the more the input values match the rule.

In Sect. 4.6.3, the explanation of the recommender is presented. It should be emphasized that in this chapter, two kinds of fuzzy IF-THEN rules are proposed: the recommendation and explanation rules (see Sect. 4.6.1). Thus, the recommender can be viewed as a special case of the neuro-fuzzy system in two different ways. When applied in order to produce recommendations, the neuro-fuzzy architecture shown in Fig. 4.34 represents the recommender. During the explanation, the network portrayed in Fig. 5.2, with explanation rules (4.5), is employed (see Sect. 5.1).

The neuro-fuzzy network presented in Fig. 4.34 reflects a profile of a particular user (investor). The Gaussian membership functions of the fuzzy sets in recommendation rules (4.4) describe the past transactions of this user that correspond to the fuzzy IF-THEN rules. At the inputs of this network there are values of attributes of new candidates to the recommendation. At the output of the network, the decision "recommend" is produced if the value of the $alignmentWithStrategy$ function, expressed by Eq. (4.8), is above a certain threshold.

For every user's network, as illustrated in Fig. 4.34, the visualizations proposed in Sect. 4.6.2, e.g. Figs. 4.30, 4.32, can be presented for every inputs (candidates to the recommendation). The neuro-fuzzy networks, for particular users, differ in the Gaussian membership functions—that correspond to the past transactions of the users. But the same input values can be processed by every network (of course, with different decisions at the output). This can be realized in parallel.

Thus, we can say that the recommender has a form of a large neuro-fuzzy network, composed of many smaller networks (equal to the number of the users) of the architecture portrayed in Fig. 4.34, with parallel processing. The number of neurons corresponding to the number of rules equals to the number of the past transactions of the user.

Such a network is not static but rather dynamic, adaptive. This means that the neuro-fuzzy recommender can change its architecture. This can be realized by creating a new network (profile) for a new user or remove a network representing a non-active user that is not interesting in the recommendations anymore.

It should be emphasized that a great advantage of the recommender, proposed in this chapter, is the elimination of the "cold start" problem. As mentioned in Sect. 1.3, it occurs when a system cannot draw any inferences for users or items about which it has not collected enough information. However, it refers to the "collaborative filtering", while the one-class recommender is considered within the "content-based techniques", this problem can also appear. This could concern a situation when the system produces recommendations for a new user (investor) who does not have his history of past transactions. In such a case, the recommender cannot infer any recommendation.

However, there is no "cold start" problem for the proposed neuro-fuzzy recommender. For a new user, without his history of past transactions, the recommender does offer a recommendation. This can be done based on the explanation rules that have been formulated by use of the histograms—which contain knowledge about past transactions of all other users (investors). Later, when the number of own transactions of the user increases, the recommendations for him can be produced based on his past transactions, employing the recommendation rules.

References

1. Agarwal, S., Sureka, A.: Using KNN and SVM based one-class classifier for detecting online radicalization on twitter. In: International Conference on Distributed Computing and Internet Technology (ICDCIT 2015), pp. 431–442. Springer, Berlin (2015)
2. Brown, S.J., Schwarz, C.: Do Market Participants Care about Portfolio Disclosure? Evidence from Hedge Funds' 13F Filings (2013)

3. De Mitri, C., Pascali, E.: Characterization of fuzzy topologies from neighborhood of fuzzy points. J. Math. Anal. Appl. **93**(1), 1–14 (1983)
4. Deshpande, M., Karypis, G.: Item-based top-N recommendation algorithms. ACM Trans. Inf. Syst. **22**(1), 143–177 (2004)
5. Dijkman, J.G., van Haeringen, H., de Lange, S.J.: Fuzzy numbers. J. Math. Anal. Appl. **92**(2), 301–341 (1983)
6. Dubois, D., Prade, H.: Fuzzy Sets and Systems: Theory and Applications. Academic, London (1980)
7. Ekstrand, M.D., Riedl, J.T., Konstan, J.A.: Collaborative filtering recommender systems. Found. Trends® Human–Comput. Interact. **4**(2), 81–173 (2011)
8. Ganster, M., Georgiou, D.N., Jafari, S., Moshokoa, S.P.: On some applications of fuzzy points. Appl. General Topol. **6**(2), 119–133 (2005)
9. Hinton, G.E., Roweis, S.T.: Stochastic neighbor embedding. In: Advances in Neural Information Processing Systems, vol. 15, pp. 833–840. Cambridge. MA. USA. The MIT Press (2002)
10. Imran, Q.H., Melgat, B.M.: New characterization of kernel set in fuzzy topological spaces. Int. J. Comput. Eng. Technol. (IJCET) **5**, 165–174 (2014)
11. Khan, S.S., Madden, M.G.: A survey of recent trends in one class classification. In: Coyle, L., Freyne, J. (eds.) AICS 2009, LNAI 6206, pp. 188–197. Springer, Berlin (2010)
12. Mercer, R.E., Barron, J.L., Bruen, A.A., Cheng, D.: Fuzzy points: algebra and application. Pattern Recognit. **35**(5), 1153–1166 (2002)
13. Oza, P., Patel, V.M.: One-class convolutional neural network. IEEE Signal Process. Lett. **26**(2), 277–281 (2019)
14. Pan, R., Zhou, Y., Cao, B., Liu, N.N., Lukose, R., Scholz, M., Yang, Q.: One-class collaborative filtering. In: ICDM '08: Proceedings of the 2008 Eighth IEEE International Conference on Data Mining, pp. 502–511 (2008)
15. Perera, P., Patel, V.M.: Learning deep features for one-class classification. IEEE Trans. Image Process. **28**(11), 5450–5463 (2019)
16. Powers, D.M.W.: Evaluation from precision, recall and F-measure to ROC, informedness, markedness and correlation. J. Mach. Learn. Technol. **2**(1), 37–63 (2011)
17. Richi, F., Rokach, L., Shapira, B., Kantor, P.B. (eds.): Recommender Systems Handbook. Springer, Berlin (2011)
18. Ruff, L., Vandermeulen, R., Goernitz, N., Deecke, L., Siddiqui, S.A., Binder, A., Muller, E., Kloft., M.: Deep one-class classification. In: Proceedings of the 35th International Conference on Machine Learning, pp. 4393–4402 (2018)
19. Rutkowska, D.: Neuro-fuzzy Architectures and Hybrid Learning. Physica-Verlag. Springer, Heidelberg, New York (2002)
20. Tax, D.M.J.: One Class Classification: Concept-Learning in the Absence of Counter-Examples, Ph.D. Thesis. Delft University of Technology (2001)
21. Tax, D.M.J., Duin, R.P.W.: Uniform object generation for optimizing one-class classifiers. J. Mach. Learn. Res. **2**, 155–173 (2001)
22. Van der Maaten, L.J.P., Hinton, G.E.: Visualizing data using t-SNE. J. Mach. Learn. Res. **9**, 2579–2605 (2008)
23. Zadeh, L.A.: Fuzzy sets. Inf. Control **8**(3), 338–353 (1965)
24. Zadeh, L.A.: Towards a theory of fuzzy systems. In: Kalman R.E. and deClaris N. (eds.). Aspect of Network and System Theory. Holt, Rinehart and Winston. New York (1971)

Chapter 5
Summary and Final Remarks

5.1 Summary of the Contributions and Novelties

Most of the novel contributions of the author, described in Chap. 3, have already been published in [7, 8]. In Chap. 3 these results are organized in a systematic way, structured under the names of recommenders A, B, and C. The new methodology for constructing and learning the neuro-fuzzy explainable recommendation systems and their performance is presented on the example of the MovieLens dataset.

In Chap. 4, a novel type of recommenders is introduced. This is a one-class classification (OCC) system (see e.g., [9]), viewed as a special case of a neuro-fuzzy (N-F) connectionist network, easy to be explainable. Figure 5.1 presents a general scheme that illustrates the performance of the recommender along with its explanation facility. With regard to the performance, this scheme refers to Fig. 2.2, which reflects the rules portrayed in Fig. 5.1 with the fuzzy sets and T-norm as the product operation.

Figure 5.1 can serve to explain the performance of the recommender. At the inputs there are values of attributes in the form of vector $\overline{\mathbf{a}}_i$, for $i = 1, 2, \ldots, n$, as denoted in Sect. 3.1.1. The first layer of the scheme includes functions Φ_i that transform the input values into vector $\overline{\mathbf{x}} = [\overline{x}_1, \overline{x}_2, \ldots, \overline{x}_n]^T$; see Sect. 2.3. This means that functions Φ_i, for $i = 1, 2, \ldots, n$, realize encoding of input attribute values according to the methods proposed in Chap. 3 or using other algorithms of feature encoding. Of course, it is required in the case of nominal (categorical) values of the attributes, e.g., for the MovieLens dataset applied in Chap. 3. Moreover, the methods of feature encoding, described in Chap. 3, allow transforming the multiple nominal values into one numerical value.

When only the feature encoding is considered, assuming that the same algorithm is employed for every attribute, we can use the same transformation function $\Phi = \Phi_1 = \Phi_2 = \cdots = \Phi_n$. However, this part of the scheme may also include the preprocessing if necessary. For example, some of the attribute values should be normalized. Thus,

T. Rutkowski, *Explainable Artificial Intelligence Based on Neuro-Fuzzy Modeling with Applications in Finance*, Studies in Computational Intelligence 964,
https://doi.org/10.1007/978-3-030-75521-8_5

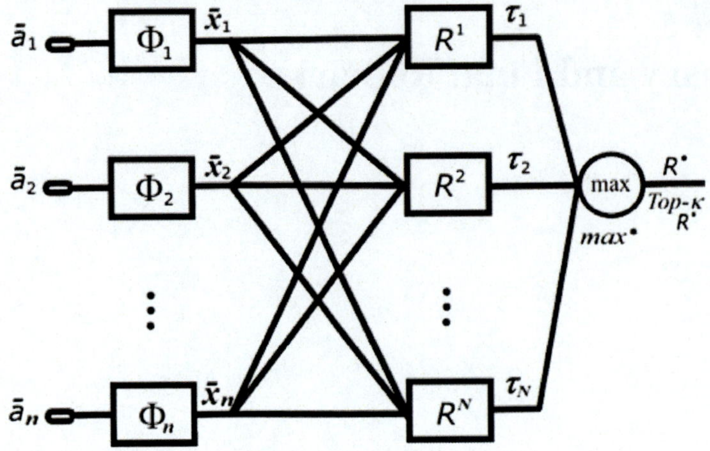

Fig. 5.1 General scheme for explanation of the recommenders' performance

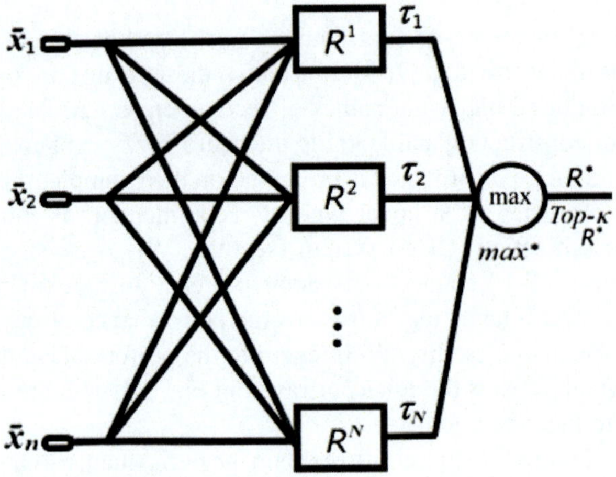

Fig. 5.2 Scheme for the explainable recommender proposed in Chap. 4

in general, the transformation functions Φ_i, for $i = 1, 2, \ldots, n$, differ between each other.

Output values of the Φ_i blocks, for $i = 1, 2, \ldots, n$, are numerical and can be directly used by the recommender. Thus, for the real-live example of the dataset prepared for investment advisers, in Chap. 4, we can explain the recommendations by means of the simpler scheme, as presented in Fig. 5.2.

In both schemes, shown in Figs. 5.1 and 5.2, input values are numerical. The layer with the rule blocks (the second layer in Fig. 5.1 and the first one in Fig. 5.2), with regard to the system performance, could be directly replaced by the network portrayed in Fig. 2.2, assuming that the product T-norm is applied. However, different T-norm functions are employed, e.g., the Dombi T-norm in Chap. 3, with regard to recommender C, and the T-norm used in the one-class recommender, in Chap. 4.

Therefore, the networks depicted in Fig. 2.2 should be modified in an appropriate way in these cases.

In spite of the fact that different T-norm functions can be applied, and other modifications introduced, e.g., weights (see Chap. 3), the schemes presented in Figs. 5.1 and 5.2 are suitable for the explanations. The main issue concerning the performance of the recommenders is to check how the numerical input values match the rules. The value of τ_k expresses the degree of activation of rule R^k, for $k = 1, 2, \ldots, N$; see Sect. 2.2. The higher this value is, the better the input values match the rule. Therefore, then in the scheme, the max block realizes the maximum function, in two ways, determining: (1) the maximum of the values τ_k, for $k = 1, 2, \ldots, N$, and (2) more than one top values of τ_k. The former case refers to the max function while the latter—to the operation denoted as $max^\star$ that produces more than one output value. Speaking more formally, the $max^\star$ operation takes τ_k, for $k = 1, 2, \ldots, N$, as arguments and determines the first max, second max, and so on, until the required number of recommendations, r, where $1 < r < N$.

The number of recommendations (items recommended to a user) can be infer from values of τ_k, for $k = 1, 2, \ldots, N$. A threshold value can be set with regard to the rule firing level, τ_k. Then, the input values (numerical attribute values) that match the rules, R^k, with values of τ_k above this threshold, are considered as the recommendations.

For a deeper explanation, the rules with the highest value of τ_k can be presented to a user, along with membership functions of the fuzzy sets in these rules. It is easy to show that the particular input (attribute) values match the membership functions with certain degrees, sufficient to activate these rules. Thus, the full explanation of the recommender performance is based on the scheme illustrated in Fig. 5.1, or Fig. 5.2, and the rules with the membership functions of fuzzy sets that have semantic meaning, as shown in Figs. 4.16, 4.20, 4.24, and in Appendix E.; e.g., Low, Medium, High. The explanation rules (with such fuzzy sets) are formulated in Sect. 4.5 and Appendix C, for 3 and 21 attributes, respectively.

With regard to recommenders A, B, and C, described in Chap. 3, the scheme depicted in Fig. 5.3 should be used for the explanation of the recommendations proposed by the systems. However, the full scheme, portrayed in Fig. 5.1, can also be applied. The important difference is that the recommenders presented in Chap. 3 are multi-class classifiers while the recommender described in Chap. 4 is the OCC system. Therefore, the interpretation of the system results is different in the case when the rule base contains IF-THEN rules with different conclusion parts, associated with particular classes.

The OCC systems, instead of classifying data items to different classes, should distinguish the data items belonging to one class from others that do not belong to this class. This seems to be a binary (two-class) classification, but the problem is totally different because there are no data items (or there are very seldom) from another class. Therefore, such a problem is often viewed as outliers recognition; see e.g., [4, 9]. The outliers are treated as data out of the class. There are also methods that try to find a boundary around the data in the one class. The term one-class classification (OOC) was introduced in [2].

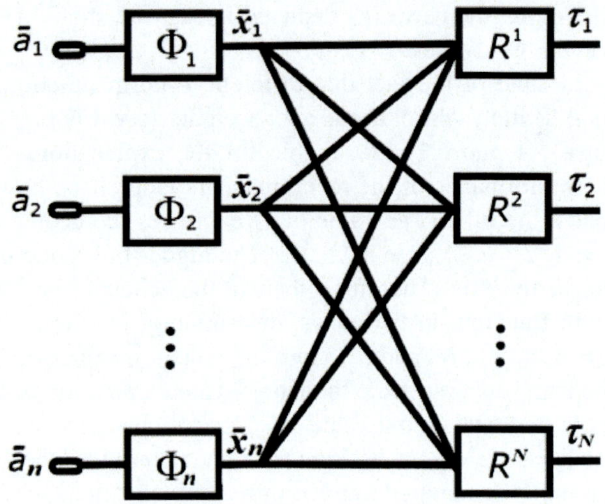

Fig. 5.3 Scheme for the explainable recommenders proposed in Chap. 3

The important contribution, with regard to the new one-class recommender, is the analysis of the histograms in order to determine the fuzzy sets, allowing to obtain fully explainable fuzzy IF-THEN rules. In particular, the idea of using the 5, 25, 50, 75 and 95th percentile (see Figs. 4.15, 4.19, and 4.23) to every histogram of the real data (for particular 21 attributes) is worth noticing. The partitioning of the attribute domains in this way produces the fuzzy sets that are easy to interpret within the explanation of recommendations generated for specific users, e.g., investors.

Another issue is the data preparation, before the visualization in the form of the histograms. This is also important for real-life data. As described in Sect. 4.3.1, the data enrichment is required to obtain the dataset when data from different sources are used. Then, features' selection proceeds the fuzzy sets' definition, as shown in Fig. 4.1.

Concerning the feature selection, in the real problem (the data used for investment advisers), and the one-class recommender, it is necessary to select the attributes taking into account users' point of view. Different algorithms employed in the case of multi-class classification are not suitable in this application because only data items from one class are available. For the explainable recommender, the attributes that are important for users should be included and analyzed.

In Chap. 3, the WM (Wang-Mendel) method of rule generation with symmetric (Gaussian) membership functions are applied. Similar results are obtained for triangular (also symmetric) functions. It should be emphasized that the membership functions of these shapes are most often used in literature (see e.g., [1, 3, 5, 6, 10]). In Chap. 3, as described earlier, the membership functions are tuned by the learning procedure, but the symmetric shape of these functions are preserved.

As we see in Figs. 4.16, 4.20, 4.24, and E.1–E.18 in Appendix D, the fuzzy sets determined based on the analysis of the histograms of the attribute data are defined by asymmetric-shape membership functions (trapezoidal). Of course, in the multi-class classifier, it is also possible to employ fuzzy sets with asymmetric membership functions. However, there is no explanation for such a choice when the MovieLens dataset is considered.

For the recommender described in Chap. 4, also the WM method is used in order to generate the fuzzy IF-THEN rules based on fuzzy sets defined within the attribute domain. However, instead of Gaussian or triangular membership functions, as shown in Fig. 2.3, the fuzzy sets portrayed in Figs. 4.16, 4.20, 4.24, and E.1–E.18 in Appendix E, are applied. In this way, the rules presented in Sect. 4.5 and Appendix C have been formulated.

Although it is mentioned in [10] that different membership functions can be employed, the analysis of the real-life data in order to define the membership functions illustrated in Chap. 4 is an important contribution. This approach, with regard to the one-class recommender, can be viewed as a modification of the WM method in the direction not only for rule generation but also as an explanation facility. The particular regions of the attribute space, obtained by use of the 5, 25, 50, 75 and 95th percentile (see Figs. 4.15, 4.19, and 4.23), give a good visualizations of the explanation.

From the explainability point of view, the approach of knowledge acquisition based on data analysis is much more suitable than any machine learning (ML) data-driven algorithm. The approach of the rule generation proposed in Chap. 4, and applied for the real-life data, based on the histograms, ought to be viewed within the framework of XAI (Explainable AI). The particular regions of the attribute space, in the explanation facility of the recommender, can also be considered with regard to the EML (Explainable ML) that is an important branch of AI within the XAI.

The difference between the EML and XAI mainly refers to the methods: data-driven in ML and knowledge-driven in AI. Shortly speaking, the EML means adding explainability to data-driven algorithms, while XAI concerns fully explainable approaches. We can say that the former is the area of research trying to introduce explainability to "black box" methods (e.g., neural networks, deep learning), and the latter is associated with methods developed as "white box" models that are interpretable, with explanation facilities.

5.2 Future Research

Every recommender, presented in Chaps. 3 and 4, is a type of the content-based (see Sect. 1.3) recommendation system. Such systems select items (objects) based on the correlation between the content of the items (characterized by their attributes) and the user's preferences.This is in contrary to the collaborative filtering systems that choose items based on the correlation between people with similar preferences. As mentioned in Sect. 1.3, the content-based approach has its great advantage of not requiring data on other users but is vulnerable to overfitting. This means that the recommender is good at fitting to the data used in order to generate the rules. However, if there is not enough data, the recommender cannot be so good at recommending new products to customers (items not applied in the process of the rule generation).

Therefore, in future research, the hybrid approach will be developed, with the goal resulting in the most accurate list of recommendations with a more precise specification of the users' profiles. The data concerning other users, investors—with regard to the datasets considered in Chap. 4, that are included in the process of rule generation (based on the histograms) can be helpful in this task.

As mentioned in Sect. 4.7, with regard to the "cold start" problem, the neuro-fuzzy one-class recommender can use the explanation network in order to produce recommendations for new users (without the knowledge of their past transactions), and in this way omit this disadvantage. Thus, in future research, the recommender that employs both the recommendation and explanation rules depending on the knowledge about users' profiles will be developed.

The feedback from users will be applied in order to update their profiles, represented by the neuro-fuzzy networks. For example, some neurons (corresponding to rules) may be removed, e.g., the oldest past transactions or those evaluated as less profitable by the users. On the other hand, the weights of the neurons could be enhanced based on their high assessment with regard to how beneficial they are for them.

The neuro-fuzzy networks that reflect users' profiles are dynamic and adaptive, also updated when new investments appear. Their assessment by different users can be incorporated into the system. Thus, future research will concern the recommender working in the framework of "human-in-the-loop".

5.3 Author's Contribution

In this section, a short list of the novelties introduced in this book is presented as follows:

- A new method of multidimensional data visualization in the 2D space have been proposed and applied (see Sect. 4.6.2)
- A new similarity measure based on the rule firing level, defined by the author, have been introduced and applied (see Sects. 4.6.1 and 4.6.2)
- The $alignmentWithStrategy$ function has been defined and applied; see Eq. (4.8) and Sect. 4.6.3
- A new method of generating recommendations based on the recommendations rules, proposed by the author, have been introduced in this book (see Sects. 4.6.1, 4.6.2, and 4.6.3)
- A new method of explanation of the recommender's performance, based on the explanation rules and the method of multidimensional data visualization in the 2D space, both proposed by the author, have been introduced in this book (see Sects. 4.6.1 and 4.6.3)
- The recommendation and explanation fuzzy IF-THEN rules have been proposed, formulated, and applied (see Sects. 4.5, 4.6.1, and Appendix C)

- Membership functions of fuzzy sets in the explanation rules have been defined based on the histograms (see Sect. 4.4 and Appendices D, E)
- The problem definition (see Sect. 4.2), selection of the attributes (see Sect. 4.3), the proposition of the schemes of generating and explaining recommendations for investment advisers (see Figs. 4.1 and 4.2), have been introduced
- A new one-class recommender considered within the framework of a large neuro-fuzzy network composed of the smaller networks representing users' profiles, with parallel processing, fully interpretable and explainable, have been proposed, constructed, and successfully applied (see Chap. 4).

All the listed above main novelties concern Chap. 4 that includes new, not published yet, author's contributions. As mentioned earlier, the propositions of novel recommenders—already published by the author—are described in Chap. 3. However, those original results have been presented in a different way in this book.

References

1. Jang, J.S.R.: ANFIS: adaptive-network-based fuzzy inference system. IEEE Trans. Syst. Man Cybern. **23**(3), 665–685 (1993)
2. Moya, M., Koch, M., Hostetler, L.: One-class classifier networks for target recognition applications. In: Proceedings of the World Congress on Neural Networks, Portland, OR. International Neural Network Society, INNS, pp. 797–801 (1993)
3. Nauck, D., Kruse, R.: How the learning of rule weights affects the interpretability of fuzzy systems. In: Proceedings of the IEEE International Conference on Fuzzy Systems 1998 (FUZZ-IEEE'98), vol. 2, pp. 1235–1240 (1998)
4. Ritter, G., Gallegos, M.: Outliers in statistical pattern recognition and an application to automatic chromosome classification. Pattern Recognit. Lett. **18**(6), 525–539 (1997)
5. Rutkowska, D.: Neuro-Fuzzy Architectures and Hybrid Learning. Physica-Verlag. Springer, Heidelberg, New York (2002)
6. Rutkowski, L.: Computational Intelligence: Methods and Techniques. Springer, Berlin (2008)
7. Rutkowski, T., Romanowski, J., Woldan, P., Staszewski, P., Nielek, R., Rutkowski, L.: A content-based recommendation system using neuro-fuzzy approach. In: 2018 IEEE International Conference on Fuzzy Systems (FUZZ-IEEE), pp. 1–8. IEEE (2018)
8. Rutkowski, T., Łapa, K., Jaworski, M., Nielek, R., Rutkowska, D.: On explainable flexible fuzzy recommender and its performance evaluation using the Akaike information criterion. In: International Conference on Neural Information Processing (ICONIP 2019), pp. 717–724. Springer, Berlin (2019)
9. Tax, D.M.J.: One Class Classification: Concept-Learning in the Absence of Counter-Examples, Ph.D. Thesis. Delft University of Technology (2001)
10. Wang, L.-X., Mendel, J.M.: Generating fuzzy rules by learning from examples. IEEE Trans. Syst. Man Cybern. **22**(6), 1414–1427 (1992)

Appendix A
Description of Attributes—Full Version

In addition to the three attributes presented in Sect. 4.3.2, the following ones are also considered as the features describing the recommendation problem. The description of the features comes from (www.investopedia.com):

enterprisevalue:
Enterprise Value (EV) is a measure of a company's total value, often used as a more comprehensive alternative to equity market capitalization. EV includes in its calculation the market capitalization of a company but also short-term and long-term debt as well as any cash on the company's balance sheet. Enterprise value is a popular metric used to value a company for a potential takeover.

pricetoearnings:
The Price-to-Earnings ratio (P/E ratio) is the ratio for valuing a company that measures its current share price relative to its Earnings Per-Share (EPS). The price-to-earnings ratio is also sometimes known as the price multiple or the earnings multiple.

evtorevenue:
The Enterprise Value-to-Revenue multiple (EV/R) is a measure of the value of a stock that compares a company's enterprise value to its revenue. EV/R is one of several fundamental indicators that investors use to determine whether a stock is priced fairly. The EV/R multiple is also often used to determine a company's valuation in the case of a potential acquisition. It's also called the enterprise value-to-sales multiple.

profitmargin:
Businesses and individuals across the globe perform for-profit economic activities with an aim to generate profits. However, absolute numbers—like X million worth of gross sales, Y thousand business expenses or $\$Z$ earnings—fail to provide a clear and realistic picture of a business' profitability and performance. Several different quantitative measures are used to compute the gains (or losses) a business generates, which make it easier to assess the performance of a business over different time periods, or compare it against competitors.

operatingmargin:
The operating margin measures how much profit a company makes on a dollar of sales, after paying for variable costs of production, such as wages and raw materials, but before paying interest or tax. It is calculated by dividing a company's operating profit by its net sales.

T. Rutkowski, *Explainable Artificial Intelligence Based on Neuro-Fuzzy Modeling with Applications in Finance*, Studies in Computational Intelligence 964,
https://doi.org/10.1007/978-3-030-75521-8

roa:
Return On Assets (ROA) is an indicator of how profitable a company is relative to its total assets. ROA gives a manager, investor, or analyst an idea as to how efficient a company's management is at using its assets to generate earnings. Return on assets is displayed as a percentage.

roe:
Return On Equity (ROE) is a measure of financial performance calculated by dividing net income by shareholders' equity. Because shareholders' equity is equal to a company's assets minus its debt, ROE could be thought of as the return on net assets.

totalrevenue:
Revenue is the amount of money that a company actually receives during a specific period, including discounts and deductions for returned merchandise. It is the top line or gross income figure from which costs are subtracted to determine net income.

revenuegrowth:
Quarterly revenue growth is an increase in a company's sales when compared to a previous quarter's revenue performance. The current quarter's sales figure can be compared on a year-over-year basis or sequentially. This helps to give analysts, investors and additional stakeholders an idea of how much a company's sales are increasing over time.

totalgrossprofit:
Gross profit is the profit a company makes after deducting the costs associated with making and selling its products, or the costs associated with providing its services. Gross profit will appear on a company's income statement and can be calculated by subtracting the Cost Of Goods Sold (COGS) from revenue (sales). These figures can be found on a company's income statement.

ebitda:
EBITDA, or Earnings Before Interest, Taxes, Depreciation and Amortization, is a measure of a company's overall financial performance and is used as an alternative to simple earnings or net income in some circumstances. However, the EBITDA can be misleading because it strips out the cost of capital investments like property, plant, and equipment.

debt:
Debt is an amount of money borrowed by one party from another. Debt is used by many corporations and individuals as a method of making large purchases that they could not afford under normal circumstances. A debt arrangement gives the borrowing party permission to borrow money under the condition that it is to be paid back at a later date, usually with interest.

debttoequity:
The Debt-to-Equity (D/E) ratio is calculated by dividing a company's total liabilities by its shareholder equity. These numbers are available on the balance sheet of a company's financial statements.

freecashflow:
Free cash flow represents the cash a company generates after cash outflows to support operations and maintain its capital assets. Unlike earnings or net income, free cash flow is a measure of profitability that excludes the non-cash expenses of the income statement and includes spending on equipment and assets as well as changes in working capital.

52weekhigh:
A 52-week high is the highest price that a stock has traded at during the previous year. It is a technical indicator used by some traders and investors who view the 52-week high as an important factor in determining a stock's current value and predicting future price movement. As a stock trades within its 52-week price range (the range that exists between the 52-week low and the 52-week high), these investors may show increased interest as price nears the high.

52weeklow:
A 52-week low is the lowest price that a stock has traded at during the previous year. It is a technical indicator used by some traders and investors who view the 52-week low as an important factor in determining a stock's current value and predicting future price movement. As a stock trades within its 52-week price range (the range that exists between the 52-week low and the 52-week high), these investors may show increased interest as price nears the low.

dividendyield:
A stock's dividend yield is expressed as an annual percentage and is calculated as the company's annual cash dividend per share divided by the current price of the stock. The dividend yield is found in the stock quotes of dividend-paying companies. Investors should note that stock quotes record the per share dollar amount of a company's latest quarterly declared dividend. This quarterly dollar amount is annualized and compared to the current stock price to generate the per annum dividend yield, which represents an expected return.

dividendrate:
The dividend rate is the total expected dividend payments from an investment, fund or portfolio expressed on an annualized basis plus any additional non-recurring dividends that an investor may receive during that period. Depending on the company's preferences and strategy, the dividend rate can be fixed or adjustable.

divpayoutratio:
The dividend payout ratio is the ratio of the total amount of dividends paid out to shareholders relative to the net income of the company. It is the percentage of earnings paid to shareholders in dividends. The amount that is not paid to shareholders is retained by the company to pay off debt or to reinvest in core operations. It is sometimes simply referred to as the 'payout ratio.'

epsgrowth:
Earnings Per Share (EPS) is the portion of a company's profit allocated to each share of common stock. Earnings per share serve as an indicator of a company's profitability. It is common for a company to report EPS that is adjusted for extraordinary items and potential share dilution.

totalcurrentassets:
Current assets represent all the assets of a company that are expected to be conveniently sold, consumed, utilized or exhausted through the standard business operations, which can lead to their conversion to a cash value over the next one year period. Since current assets is a standard item appearing in the balance sheet, the time horizon represents one year from the date shown in the heading of the company's balance sheet.

Appendix B
Fuzzy IF-THEN Rules

The following rules have been formulated for the simplified version of the recommendation problem (characterized by three attributes presented in Sect. 4.3.2):

1. IF Current Ratio is Very High AND Enterprise To EBIDTA is Very High AND Price To Book is Very High THEN decision is YES
2. IF Current Ratio is Medium AND Enterprise To EBIDTA is Very High AND Price To Book is Very High THEN decision is YES
3. IF Current Ratio is Very High AND Enterprise To EBIDTA is Very Low AND Price To Book is Very High THEN decision is YES
4. IF Current Ratio is High AND Enterprise To EBIDTA is Very High AND Price To Book is Very High THEN decision is YES
5. IF Current Ratio is Low AND Enterprise To EBIDTA is Very High AND Price To Book is Very High THEN decision is YES
6. IF Current Ratio is Medium AND Enterprise To EBIDTA is High AND Price To Book is Very High THEN decision is YES
7. IF Current Ratio is Very High AND Enterprise To EBIDTA is High AND Price To Book is Very High THEN decision is YES
8. IF Current Ratio is Very High AND Enterprise To EBIDTA is Very High AND Price To Book is High THEN decision is YES
9. IF Current Ratio is Low AND Enterprise To EBIDTA is Very High AND Price To Book is High THEN decision is YES
10. IF Current Ratio is Low AND Enterprise To EBIDTA is Medium AND Price To Book is Very High THEN decision is YES
11. IF Current Ratio is Low AND Enterprise To EBIDTA is High AND Price To Book is Very High THEN decision is YES
12. IF Current Ratio is High AND Enterprise To EBIDTA is Very Low AND Price To Book is Very High THEN decision is YES
13. IF Current Ratio is Medium AND Enterprise To EBIDTA is Very High AND Price To Book is High THEN decision is YES
14. IF Current Ratio is Low AND Enterprise To EBIDTA is Very High AND Price To Book is Medium THEN decision is YES
15. IF Current Ratio is Very Low AND Enterprise To EBIDTA is Very High AND Price To Book is Medium THEN decision is YES
16. IF Current Ratio is High AND Enterprise To EBIDTA is High AND Price To Book is Very High THEN decision is YES

T. Rutkowski, *Explainable Artificial Intelligence Based on Neuro-Fuzzy Modeling with Applications in Finance*, Studies in Computational Intelligence 964,
https://doi.org/10.1007/978-3-030-75521-8

Appendix B
Fuzzy IF-THEN Rules

The following rules have been formulated for the simplified version of the recommendation problem (characterized by three attributes presented in Sect. 4.3.2):

1. IF Current Ratio is Very High AND Enterprise To EBIDTA is Very High AND Price To Book is Very High THEN decision is YES
2. IF Current Ratio is Medium AND Enterprise To EBIDTA is Very High AND Price To Book is Very High THEN decision is YES
3. IF Current Ratio is Very High AND Enterprise To EBIDTA is Very Low AND Price To Book is Very High THEN decision is YES
4. IF Current Ratio is High AND Enterprise To EBIDTA is Very High AND Price To Book is Very High THEN decision is YES
5. IF Current Ratio is Low AND Enterprise To EBIDTA is Very High AND Price To Book is Very High THEN decision is YES
6. IF Current Ratio is Medium AND Enterprise To EBIDTA is High AND Price To Book is Very High THEN decision is YES
7. IF Current Ratio is Very High AND Enterprise To EBIDTA is High AND Price To Book is Very High THEN decision is YES
8. IF Current Ratio is Very High AND Enterprise To EBIDTA is Very High AND Price To Book is High THEN decision is YES
9. IF Current Ratio is Low AND Enterprise To EBIDTA is Very High AND Price To Book is High THEN decision is YES
10. IF Current Ratio is Low AND Enterprise To EBIDTA is Medium AND Price To Book is Very High THEN decision is YES
11. IF Current Ratio is Low AND Enterprise To EBIDTA is High AND Price To Book is Very High THEN decision is YES
12. IF Current Ratio is High AND Enterprise To EBIDTA is Very Low AND Price To Book is Very High THEN decision is YES
13. IF Current Ratio is Medium AND Enterprise To EBIDTA is Very High AND Price To Book is High THEN decision is YES
14. IF Current Ratio is Low AND Enterprise To EBIDTA is Very High AND Price To Book is Medium THEN decision is YES
15. IF Current Ratio is Very Low AND Enterprise To EBIDTA is Very High AND Price To Book is Medium THEN decision is YES
16. IF Current Ratio is High AND Enterprise To EBIDTA is High AND Price To Book is Very High THEN decision is YES

T. Rutkowski, *Explainable Artificial Intelligence Based on Neuro-Fuzzy Modeling with Applications in Finance*, Studies in Computational Intelligence 964,
https://doi.org/10.1007/978-3-030-75521-8

17. IF Current Ratio is Very High AND Enterprise To EBIDTA is Very Low AND Price To Book is High THEN decision is YES
18. IF Current Ratio is Low AND Enterprise To EBIDTA is High AND Price To Book is High THEN decision is YES
19. IF Current Ratio is Low AND Enterprise To EBIDTA is Medium AND Price To Book is High THEN decision is YES
20. IF Current Ratio is Low AND Enterprise To EBIDTA is Very Low AND Price To Book is Very High THEN decision is YES
21. IF Current Ratio is High AND Enterprise To EBIDTA is Very High AND Price To Book is Medium THEN decision is YES
22. IF Current Ratio is High AND Enterprise To EBIDTA is Medium AND Price To Book is Very High THEN decision is YES
23. IF Current Ratio is Medium AND Enterprise To EBIDTA is Very Low AND Price To Book is Very High THEN decision is YES
24. IF Current Ratio is Very High AND Enterprise To EBIDTA is Very High AND Price To Book is Low THEN decision is YES
25. IF Current Ratio is Very High AND Enterprise To EBIDTA is High AND Price To Book is High THEN decision is YES
26. IF Current Ratio is Medium AND Enterprise To EBIDTA is High AND Price To Book is High THEN decision is YES
27. IF Current Ratio is Very High AND Enterprise To EBIDTA is Medium AND Price To Book is Very High THEN decision is YES
28. IF Current Ratio is Very High AND Enterprise To EBIDTA is Very Low AND Price To Book is Low THEN decision is YES
29. IF Current Ratio is Very Low AND Enterprise To EBIDTA is Very High AND Price To Book is Low THEN decision is YES
30. IF Current Ratio is Medium AND Enterprise To EBIDTA is Medium AND Price To Book is Very High THEN decision is YES
31. IF Current Ratio is Low AND Enterprise To EBIDTA is Very Low AND Price To Book is Low THEN decision is YES
32. IF Current Ratio is Very Low AND Enterprise To EBIDTA is Very High AND Price To Book is Very High THEN decision is YES
33. IF Current Ratio is Medium AND Enterprise To EBIDTA is Very High AND Price To Book is Medium THEN decision is YES
34. IF Current Ratio is Very High AND Enterprise To EBIDTA is Very High AND Price To Book is Medium THEN decision is YES
35. IF Current Ratio is Medium AND Enterprise To EBIDTA is Low AND Price To Book is Medium THEN decision is YES
36. IF Current Ratio is Low AND Enterprise To EBIDTA is Low AND Price To Book is High THEN decision is YES
37. IF Current Ratio is Very High AND Enterprise To EBIDTA is Low AND Price To Book is Low THEN decision is YES
38. IF Current Ratio is Very Low AND Enterprise To EBIDTA is High AND Price To Book is Medium THEN decision is YES
39. IF Current Ratio is Very High AND Enterprise To EBIDTA is Medium AND Price To Book is High THEN decision is YES
40. IF Current Ratio is High AND Enterprise To EBIDTA is Very Low AND Price To Book is High THEN decision is YES
41. IF Current Ratio is Very Low AND Enterprise To EBIDTA is Medium AND Price To Book is Medium THEN decision is YES
42. IF Current Ratio is Very High AND Enterprise To EBIDTA is Medium AND Price To Book is Medium THEN decision is YES
43. IF Current Ratio is Very High AND Enterprise To EBIDTA is Very Low AND Price To Book is Medium THEN decision is YES

44. IF Current Ratio is High AND Enterprise To EBIDTA is Very High AND Price To Book is High THEN decision is YES
45. IF Current Ratio is Very Low AND Enterprise To EBIDTA is High AND Price To Book is Low THEN decision is YES
46. IF Current Ratio is Very Low AND Enterprise To EBIDTA is Low AND Price To Book is Medium THEN decision is YES
47. IF Current Ratio is Low AND Enterprise To EBIDTA is Very Low AND Price To Book is Medium THEN decision is YES
48. IF Current Ratio is Very High AND Enterprise To EBIDTA is Low AND Price To Book is Medium THEN decision is YES
49. IF Current Ratio is Very Low AND Enterprise To EBIDTA is Very Low AND Price To Book is Medium THEN decision is YES
50. IF Current Ratio is Low AND Enterprise To EBIDTA is Low AND Price To Book is Very High THEN decision is YES
51. IF Current Ratio is Very Low AND Enterprise To EBIDTA is High AND Price To Book is High THEN decision is YES
52. IF Current Ratio is Medium AND Enterprise To EBIDTA is Medium AND Price To Book is Low THEN decision is YES
53. IF Current Ratio is Medium AND Enterprise To EBIDTA is High AND Price To Book is Medium THEN decision is YES
54. IF Current Ratio is High AND Enterprise To EBIDTA is Medium AND Price To Book is Medium THEN decision is YES
55. IF Current Ratio is High AND Enterprise To EBIDTA is Very High AND Price To Book is Low THEN decision is YES
56. IF Current Ratio is Medium AND Enterprise To EBIDTA is Very Low AND Price To Book is Low THEN decision is YES
57. IF Current Ratio is Low AND Enterprise To EBIDTA is Low AND Price To Book is Medium THEN decision is YES
58. IF Current Ratio is Medium AND Enterprise To EBIDTA is Medium AND Price To Book is High THEN decision is YES
59. IF Current Ratio is Medium AND Enterprise To EBIDTA is Low AND Price To Book is Very High THEN decision is YES
60. IF Current Ratio is Low AND Enterprise To EBIDTA is Medium AND Price To Book is Medium THEN decision is YES
61. IF Current Ratio is Low AND Enterprise To EBIDTA is Low AND Price To Book is Low THEN decision is YES
62. IF Current Ratio is Very High AND Enterprise To EBIDTA is Medium AND Price To Book is Low THEN decision is YES
63. IF Current Ratio is Medium AND Enterprise To EBIDTA is Very Low AND Price To Book is Medium THEN decision is YES
64. IF Current Ratio is High AND Enterprise To EBIDTA is High AND Price To Book is High THEN decision is YES
65. IF Current Ratio is High AND Enterprise To EBIDTA is High AND Price To Book is Medium THEN decision is YES
66. IF Current Ratio is High AND Enterprise To EBIDTA is Low AND Price To Book is Medium THEN decision is YES
67. IF Current Ratio is Low AND Enterprise To EBIDTA is Very High AND Price To Book is Low THEN decision is YES
68. IF Current Ratio is Medium AND Enterprise To EBIDTA is High AND Price To Book is Low THEN decision is YES
69. IF Current Ratio is Very High AND Enterprise To EBIDTA is High AND Price To Book is Low THEN decision is YES
70. IF Current Ratio is Medium AND Enterprise To EBIDTA is Medium AND Price To Book is Medium THEN decision is YES

71. IF Current Ratio is Very High AND Enterprise To EBIDTA is Low AND Price To Book is Very High THEN decision is YES
72. IF Current Ratio is Very Low AND Enterprise To EBIDTA is Medium AND Price To Book is Very High THEN decision is YES
73. IF Current Ratio is High AND Enterprise To EBIDTA is Medium AND Price To Book is High THEN decision is YES
74. IF Current Ratio is Medium AND Enterprise To EBIDTA is Very Low AND Price To Book is High THEN decision is YES
75. IF Current Ratio is Very High AND Enterprise To EBIDTA is Low AND Price To Book is High THEN decision is YES
76. IF Current Ratio is High AND Enterprise To EBIDTA is Low AND Price To Book is Low THEN decision is YES
77. IF Current Ratio is Very Low AND Enterprise To EBIDTA is Very Low AND Price To Book is Low THEN decision is YES
78. IF Current Ratio is Medium AND Enterprise To EBIDTA is Low AND Price To Book is High THEN decision is YES
79. IF Current Ratio is Low AND Enterprise To EBIDTA is High AND Price To Book is Medium THEN decision is YES
80. IF Current Ratio is Low AND Enterprise To EBIDTA is Medium AND Price To Book is Low THEN decision is YES
81. IF Current Ratio is High AND Enterprise To EBIDTA is Low AND Price To Book is Very High THEN decision is YES
82. IF Current Ratio is Very Low AND Enterprise To EBIDTA is Low AND Price To Book is Low THEN decision is YES
83. IF Current Ratio is Medium AND Enterprise To EBIDTA is Low AND Price To Book is Low THEN decision is YES
84. IF Current Ratio is Very Low AND Enterprise To EBIDTA is Low AND Price To Book is Very High THEN decision is YES
85. IF Current Ratio is High AND Enterprise To EBIDTA is Low AND Price To Book is High THEN decision is YES
86. IF Current Ratio is Very High AND Enterprise To EBIDTA is High AND Price To Book is Medium THEN decision is YES
87. IF Current Ratio is Very Low AND Enterprise To EBIDTA is Very High AND Price To Book is High THEN decision is YES
88. IF Current Ratio is Very Low AND Enterprise To EBIDTA is High AND Price To Book is Very High THEN decision is YES
89. IF Current Ratio is High AND Enterprise To EBIDTA is High AND Price To Book is Low THEN decision is YES
90. IF Current Ratio is High AND Enterprise To EBIDTA is Medium AND Price To Book is Low THEN decision is YES
91. IF Current Ratio is Medium AND Enterprise To EBIDTA is Very High AND Price To Book is Low THEN decision is YES
92. IF Current Ratio is High AND Enterprise To EBIDTA is Very Low AND Price To Book is Medium THEN decision is YES
93. IF Current Ratio is Very Low AND Enterprise To EBIDTA is Very Low AND Price To Book is High THEN decision is YES
94. IF Current Ratio is Very Low AND Enterprise To EBIDTA is Very Low AND Price To Book is Very High THEN decision is YES
95. IF Current Ratio is High AND Enterprise To EBIDTA is Very Low AND Price To Book is Low THEN decision is YES
96. IF Current Ratio is Very Low AND Enterprise To EBIDTA is Low AND Price To Book is High THEN decision is YES
97. IF Current Ratio is Very Low AND Enterprise To EBIDTA is Medium AND Price To Book is High THEN decision is YES

98. IF Current Ratio is Very Low AND Enterprise To EBIDTA is Medium AND Price To Book is Low THEN decision is YES
99. IF Current Ratio is Low AND Enterprise To EBIDTA is Very Low AND Price To Book is High THEN decision is YES
100. IF Current Ratio is Low AND Enterprise To EBIDTA is High AND Price To Book is Low THEN decision is YES.

Appendix C
Fuzzy Rules - Full Version

The following rules have been generated by use of all the attributes presented in Sect. 4.3.2 and Appendix A.

1. IF currentRatio IS Low AND earningsPerShareGrowth IS Low AND revenueGrowth IS Low AND currentAssets IS Low AND profitMargins IS Very low AND debtToEquity IS Low AND dividendYield IS Very low AND ebitda IS Low AND enterpriseToEbitda IS Very high AND fiftyTwoWeekHigh IS High AND fiftyTwoWeekLow IS High AND freeCashflow IS Low AND grossProfits IS Very low AND operatingMargins IS Very low AND payoutRatio IS Very low AND priceToBook IS Low AND priceToEarningsToGrowth IS Very low AND returnOnAssets IS Low AND returnOnEquity IS Low AND totalDebt IS Low AND totalDebtToCurrentAsset IS Low THEN invest.
2. IF currentRatio IS High AND earningsPerShareGrowth IS High AND revenueGrowth IS Very high AND currentAssets IS Very low AND profitMargins IS Very low AND debtToEquity IS Very low AND dividendYield IS Very low AND ebitda IS Low AND enterpriseToEbitda IS Low AND fiftyTwoWeekHigh IS High AND fiftyTwoWeekLow IS High AND freeCashflow IS Low AND grossProfits IS Very low AND operatingMargins IS Very low AND payoutRatio IS Very low AND priceToBook IS Low AND priceToEarningsToGrowth IS Very low AND returnOnAssets IS Very low AND returnOnEquity IS Very low AND totalDebt IS Very low AND totalDebtToCurrentAsset IS Very low THEN invest.
3. IF currentRatio IS Medium AND earningsPerShareGrowth IS Medium AND revenueGrowth IS High AND currentAssets IS Medium AND profitMargins IS Low AND debtToEquity IS High AND dividendYield IS Medium AND ebitda IS Medium AND enterpriseToEbitda IS Medium AND fiftyTwoWeekHigh IS High AND fiftyTwoWeekLow IS High AND freeCashflow IS Low AND grossProfits IS Medium AND operatingMargins IS Low AND payoutRatio IS Medium AND priceToBook IS Medium AND priceToEarningsToGrowth IS Medium AND returnOnAssets IS Medium AND returnOnEquity IS High AND totalDebt IS Medium AND totalDebtToCurrentAsset IS Very high THEN invest.
4. IF currentRatio IS Low AND earningsPerShareGrowth IS High AND revenueGrowth IS Medium AND currentAssets IS Very high AND profitMargins IS Medium AND debtToEquity IS Very low AND dividendYield IS Medium AND ebitda IS High AND enterpriseToEbitda IS High AND fiftyTwoWeekHigh IS Very high AND fiftyTwoWeekLow IS Very high AND freeCashflow IS Very high AND grossProfits IS High AND operatingMargins IS Medium AND payoutRatio IS High AND priceToBook IS Very low AND priceToEarningsToGrowth IS High AND returnOnAssets IS High AND returnOnEquity IS Low AND totalDebt IS Very high AND totalDebtToCurrentAsset IS Low THEN invest.

T. Rutkowski, *Explainable Artificial Intelligence Based on Neuro-Fuzzy Modeling with Applications in Finance*, Studies in Computational Intelligence 964,
https://doi.org/10.1007/978-3-030-75521-8

5. IF currentRatio IS Very high AND earningsPerShareGrowth IS Low AND revenueGrowth IS Very high AND currentAssets IS Very low AND profitMargins IS Very low AND debtToEquity IS Very low AND dividendYield IS Very low AND ebitda IS Low AND enterpriseToEbitda IS Very low AND fiftyTwoWeekHigh IS Low AND fiftyTwoWeekLow IS Very low AND freeCashflow IS Low AND grossProfits IS Very low AND operatingMargins IS Very low AND payoutRatio IS Very low AND priceToBook IS High AND priceToEarningsToGrowth IS Very low AND returnOnAssets IS Very low AND returnOnEquity IS Very low AND totalDebt IS Very low AND totalDebtToCurrentAsset IS Very low THEN invest.
6. IF currentRatio IS Medium AND earningsPerShareGrowth IS Very low AND revenueGrowth IS Very low AND currentAssets IS High AND profitMargins IS Very low AND debtToEquity IS Medium AND dividendYield IS Very low AND ebitda IS Very low AND enterpriseToEbitda IS Very high AND fiftyTwoWeekHigh IS Very high AND fiftyTwoWeekLow IS Very high AND freeCashflow IS High AND grossProfits IS Medium AND operatingMargins IS Very low AND payoutRatio IS Very low AND priceToBook IS Medium AND priceToEarningsToGrowth IS Very low AND returnOnAssets IS Low AND returnOnEquity IS Low AND totalDebt IS High AND totalDebtToCurrentAsset IS Medium THEN invest.
7. IF currentRatio IS Low AND earningsPerShareGrowth IS Low AND revenueGrowth IS Very low AND currentAssets IS Very low AND profitMargins IS Very low AND debtToEquity IS Very low AND dividendYield IS Very low AND ebitda IS Low AND enterpriseToEbitda IS Very low AND fiftyTwoWeekHigh IS High AND fiftyTwoWeekLow IS Medium AND freeCashflow IS Low AND grossProfits IS Very low AND operatingMargins IS Very low AND payoutRatio IS Very low AND priceToBook IS Very low AND priceToEarningsToGrowth IS Very low AND returnOnAssets IS Very low AND returnOnEquity IS Low AND totalDebt IS Very low AND totalDebtToCurrentAsset IS Very high THEN invest.
8. IF currentRatio IS High AND earningsPerShareGrowth IS Medium AND revenueGrowth IS High AND currentAssets IS Medium AND profitMargins IS High AND debtToEquity IS Very low AND dividendYield IS Very low AND ebitda IS Medium AND enterpriseToEbitda IS High AND fiftyTwoWeekHigh IS Medium AND fiftyTwoWeekLow IS Medium AND freeCashflow IS High AND grossProfits IS Medium AND operatingMargins IS High AND payoutRatio IS Very low AND priceToBook IS High AND priceToEarningsToGrowth IS High AND returnOnAssets IS Very high AND returnOnEquity IS High AND totalDebt IS Low AND totalDebtToCurrentAsset IS Very low THEN invest.
9. IF currentRatio IS Very high AND earningsPerShareGrowth IS Low AND revenueGrowth IS Low AND currentAssets IS Low AND profitMargins IS Very low AND debtToEquity IS Very low AND dividendYield IS Very low AND ebitda IS Very low AND enterpriseToEbitda IS Very low AND fiftyTwoWeekHigh IS Medium AND fiftyTwoWeekLow IS Low AND freeCashflow IS Low AND grossProfits IS Very low AND operatingMargins IS Very low AND payoutRatio IS Very low AND priceToBook IS High AND priceToEarningsToGrowth IS Very low AND returnOnAssets IS Very low AND returnOnEquity IS Very low AND totalDebt IS Very low AND totalDebtToCurrentAsset IS Very low THEN invest.
10. IF currentRatio IS High AND earningsPerShareGrowth IS Very high AND revenueGrowth IS High AND currentAssets IS Very high AND profitMargins IS High AND debtToEquity IS Very low AND dividendYield IS Very low AND ebitda IS Very high AND enterpriseToEbitda IS High AND fiftyTwoWeekHigh IS Very high AND fiftyTwoWeekLow IS Very high AND freeCashflow IS Very high AND grossProfits IS Very high AND operatingMargins IS High AND payoutRatio IS Very low AND priceToBook IS High AND priceToEarningsToGrowth IS High AND returnOnAssets IS High AND returnOnEquity IS High AND totalDebt IS Very high AND totalDebtToCurrentAsset IS Low THEN invest.
11. IF currentRatio IS Very high AND earningsPerShareGrowth IS Low AND revenueGrowth IS Low AND currentAssets IS Very low AND profitMargins IS Very low AND debtToEquity IS Very low AND dividendYield IS Very low AND ebitda IS Very low AND enterpriseToEbitda IS Low AND fiftyTwoWeekHigh IS High AND fiftyTwoWeekLow IS Medium AND freeCashflow IS Low AND grossProfits IS Very low AND operatingMargins IS Very low AND payoutRatio IS Very low AND priceToBook IS Low AND priceToEarningsToGrowth IS Very

low AND returnOnAssets IS Very low AND returnOnEquity IS Very low AND totalDebt IS Very low AND totalDebtToCurrentAsset IS Very low THEN invest.

12. IF currentRatio IS Very high AND earningsPerShareGrowth IS Low AND revenueGrowth IS Very low AND currentAssets IS Low AND profitMargins IS Very low AND debtToEquity IS Very low AND dividendYield IS Very low AND ebitda IS Very low AND enterpriseToEbitda IS Very low AND fiftyTwoWeekHigh IS Low AND fiftyTwoWeekLow IS Low AND freeCashflow IS Low AND grossProfits IS Very low AND operatingMargins IS Very low AND payoutRatio IS Very low AND priceToBook IS High AND priceToEarningsToGrowth IS Very low AND returnOnAssets IS Very low AND returnOnEquity IS Very low AND totalDebt IS Very low AND totalDebtToCurrentAsset IS Very low THEN invest.
13. IF currentRatio IS Very high AND earningsPerShareGrowth IS Medium AND revenueGrowth IS Very low AND currentAssets IS Low AND profitMargins IS Very low AND debtToEquity IS Very low AND dividendYield IS Very low AND ebitda IS Very low AND enterpriseToEbitda IS Low AND fiftyTwoWeekHigh IS Low AND fiftyTwoWeekLow IS Low AND freeCashflow IS Low AND grossProfits IS Very low AND operatingMargins IS Very low AND payoutRatio IS Very low AND priceToBook IS Medium AND priceToEarningsToGrowth IS Very low AND returnOnAssets IS Very low AND returnOnEquity IS Very low AND totalDebt IS Very low AND totalDebtToCurrentAsset IS Very low THEN invest.
14. IF currentRatio IS Medium AND earningsPerShareGrowth IS High AND revenueGrowth IS Medium AND currentAssets IS Medium AND profitMargins IS Low AND debtToEquity IS Medium AND dividendYield IS Very low AND ebitda IS High AND enterpriseToEbitda IS Very high AND fiftyTwoWeekHigh IS Very high AND fiftyTwoWeekLow IS Very high AND freeCashflow IS High AND grossProfits IS High AND operatingMargins IS Medium AND payoutRatio IS Very low AND priceToBook IS High AND priceToEarningsToGrowth IS Very low AND returnOnAssets IS Low AND returnOnEquity IS Low AND totalDebt IS Medium AND totalDebtToCurrentAsset IS High THEN invest.
15. IF currentRatio IS Medium AND earningsPerShareGrowth IS Medium AND revenueGrowth IS Low AND currentAssets IS Very low AND profitMargins IS Low AND debtToEquity IS Very high AND dividendYield IS Very high AND ebitda IS Low AND enterpriseToEbitda IS Medium AND fiftyTwoWeekHigh IS Very low AND fiftyTwoWeekLow IS Very low AND freeCashflow IS Low AND grossProfits IS Low AND operatingMargins IS Low AND payoutRatio IS Very low AND priceToBook IS Medium AND priceToEarningsToGrowth IS Very low AND returnOnAssets IS Low AND returnOnEquity IS Very low AND totalDebt IS Low AND totalDebtToCurrentAsset IS Very high THEN invest.
16. IF currentRatio IS High AND earningsPerShareGrowth IS Low AND revenueGrowth IS High AND currentAssets IS Low AND profitMargins IS Very low AND debtToEquity IS Very low AND dividendYield IS Very low AND ebitda IS Very low AND enterpriseToEbitda IS Very low AND fiftyTwoWeekHigh IS Low AND fiftyTwoWeekLow IS Low AND freeCashflow IS Low AND grossProfits IS Very low AND operatingMargins IS Very low AND payoutRatio IS Very low AND priceToBook IS Medium AND priceToEarningsToGrowth IS Very low AND returnOnAssets IS Very low AND returnOnEquity IS Very low AND totalDebt IS Very low AND totalDebtToCurrentAsset IS Very low THEN invest.
17. IF currentRatio IS High AND earningsPerShareGrowth IS Medium AND revenueGrowth IS Very high AND currentAssets IS Low AND profitMargins IS Very low AND debtToEquity IS High AND dividendYield IS Very low AND ebitda IS Very low AND enterpriseToEbitda IS Low AND fiftyTwoWeekHigh IS Low AND fiftyTwoWeekLow IS Low AND freeCashflow IS Low AND grossProfits IS Low AND operatingMargins IS Very low AND payoutRatio IS Very low AND priceToBook IS High AND priceToEarningsToGrowth IS Very low AND returnOnAssets IS Very low AND returnOnEquity IS Very low AND totalDebt IS Very low AND totalDebtToCurrentAsset IS High THEN invest.
18. IF currentRatio IS High AND earningsPerShareGrowth IS Medium AND revenueGrowth IS High AND currentAssets IS Medium AND profitMargins IS High AND debtToEquity IS Medium AND dividendYield IS Low AND ebitda IS Medium AND enterpriseToEbitda IS Low AND fiftyTwoWeekHigh IS Medium AND fiftyTwoWeekLow IS Medium AND

freeCashflow IS Low AND grossProfits IS Medium AND operatingMargins IS Very high AND payoutRatio IS Low AND priceToBook IS Low AND priceToEarningsToGrowth IS Medium AND returnOnAssets IS Medium AND returnOnEquity IS Medium AND totalDebt IS Low AND totalDebtToCurrentAsset IS Medium THEN invest.

19. IF currentRatio IS High AND earningsPerShareGrowth IS Low AND revenueGrowth IS Low AND currentAssets IS Medium AND profitMargins IS High AND debtToEquity IS Medium AND dividendYield IS Low AND ebitda IS High AND enterpriseToEbitda IS Low AND fiftyTwoWeekHigh IS Medium AND fiftyTwoWeekLow IS Medium AND freeCashflow IS Low AND grossProfits IS Medium AND operatingMargins IS Very high AND payoutRatio IS Low AND priceToBook IS Low AND priceToEarningsToGrowth IS Medium AND returnOnAssets IS Medium AND returnOnEquity IS Medium AND totalDebt IS Medium AND totalDebtToCurrentAsset IS Medium THEN invest.
20. IF currentRatio IS Very high AND earningsPerShareGrowth IS Low AND revenueGrowth IS Low AND currentAssets IS Low AND profitMargins IS Very low AND debtToEquity IS Very low AND dividendYield IS Very low AND ebitda IS Very low AND enterpriseToEbitda IS Very low AND fiftyTwoWeekHigh IS Medium AND fiftyTwoWeekLow IS Medium AND freeCashflow IS Low AND grossProfits IS Very low AND operatingMargins IS Very low AND payoutRatio IS Very low AND priceToBook IS High AND priceToEarningsToGrowth IS Very low AND returnOnAssets IS Very low AND returnOnEquity IS Very low AND totalDebt IS Very low AND totalDebtToCurrentAsset IS Very low THEN invest.
21. IF currentRatio IS Low AND earningsPerShareGrowth IS Medium AND revenueGrowth IS Medium AND currentAssets IS Low AND profitMargins IS Medium AND debtToEquity IS Medium AND dividendYield IS Very low AND ebitda IS High AND enterpriseToEbitda IS High AND fiftyTwoWeekHigh IS Medium AND fiftyTwoWeekLow IS Medium AND freeCashflow IS High AND grossProfits IS Medium AND operatingMargins IS High AND payoutRatio IS Very low AND priceToBook IS Low AND priceToEarningsToGrowth IS Very low AND returnOnAssets IS Medium AND returnOnEquity IS Low AND totalDebt IS Medium AND totalDebtToCurrentAsset IS Medium THEN invest.
22. IF currentRatio IS High AND earningsPerShareGrowth IS Low AND revenueGrowth IS Very low AND currentAssets IS Very low AND profitMargins IS Very low AND debtToEquity IS Very low AND dividendYield IS Very low AND ebitda IS Low AND enterpriseToEbitda IS Low AND fiftyTwoWeekHigh IS Very low AND fiftyTwoWeekLow IS Very low AND freeCashflow IS Low AND grossProfits IS Very low AND operatingMargins IS Very low AND payoutRatio IS Very low AND priceToBook IS Low AND priceToEarningsToGrowth IS Very low AND returnOnAssets IS Very low AND returnOnEquity IS Very low AND totalDebt IS Very low AND totalDebtToCurrentAsset IS Very low THEN invest.
23. IF currentRatio IS Low AND earningsPerShareGrowth IS Medium AND revenueGrowth IS High AND currentAssets IS Medium AND profitMargins IS Medium AND debtToEquity IS Very high AND dividendYield IS Very low AND ebitda IS High AND enterpriseToEbitda IS Medium AND fiftyTwoWeekHigh IS Very high AND fiftyTwoWeekLow IS Very high AND freeCashflow IS High AND grossProfits IS High AND operatingMargins IS High AND payoutRatio IS Very low AND priceToBook IS Very high AND priceToEarningsToGrowth IS Medium AND returnOnAssets IS High AND returnOnEquity IS Very high AND totalDebt IS High AND totalDebtToCurrentAsset IS High THEN invest.

Appendix D
Histograms of Attribute Values

The histograms of data for three attributes (simplified versions) are shown in Figs. 4.14, 4.18, and 4.22. Now the additional histograms to complete the full version of 21 attributes are presented in Figs. D.1, D.2, D.3, D.4, D.5, D.6, D.7, D.8, D.9, D.10, D.11, D.12, D.13, D.14, D.15, D.16, D.17 and D.18 ; see Sect. A for descriptions of these attributes.

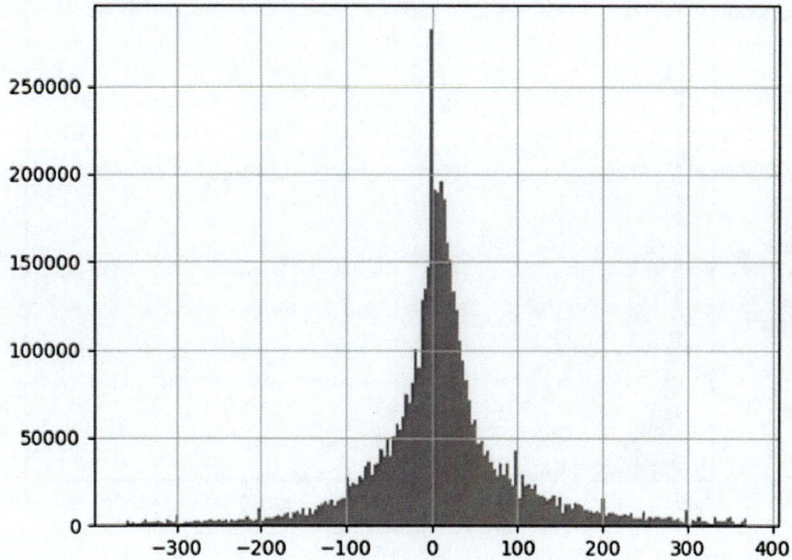

Fig. D.1 Histogram of attribute: epsgrowth

T. Rutkowski, *Explainable Artificial Intelligence Based on Neuro-Fuzzy Modeling with Applications in Finance*, Studies in Computational Intelligence 964,
https://doi.org/10.1007/978-3-030-75521-8

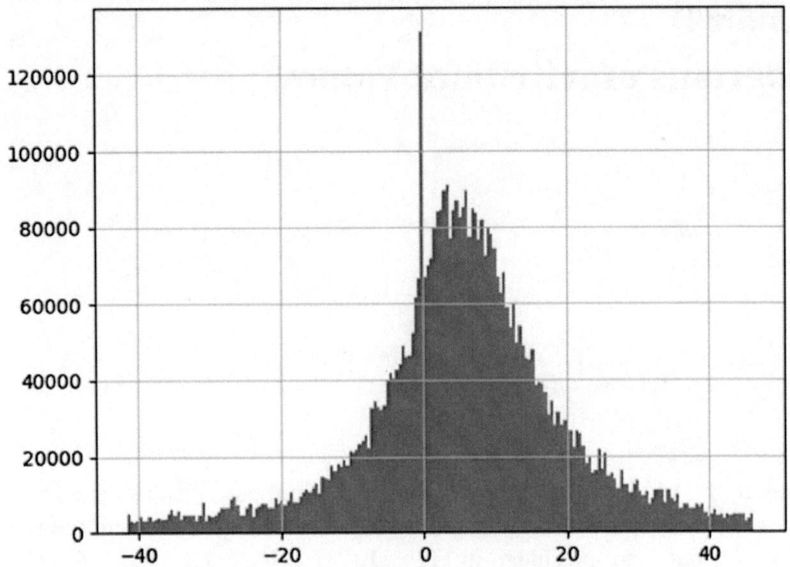

Fig. D.2 Histogram of attribute: revenuegrowth

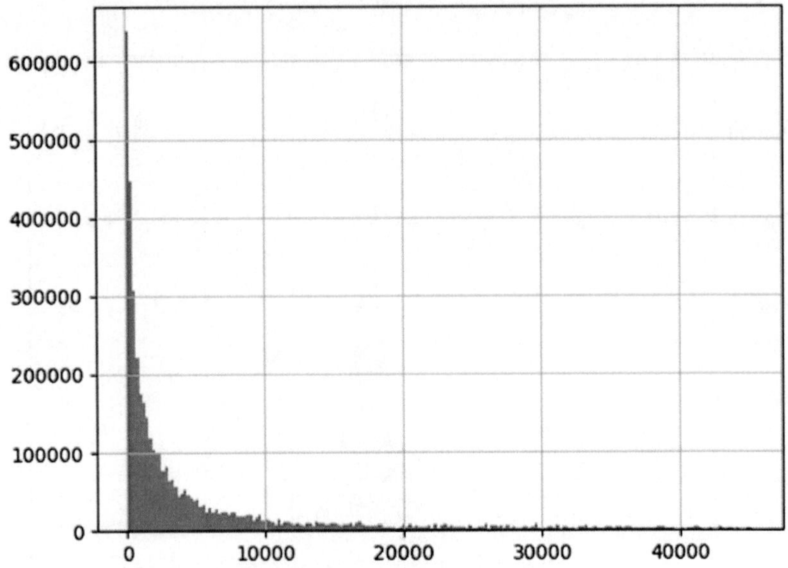

Fig. D.3 Histogram of attribute: totalcurrentassets

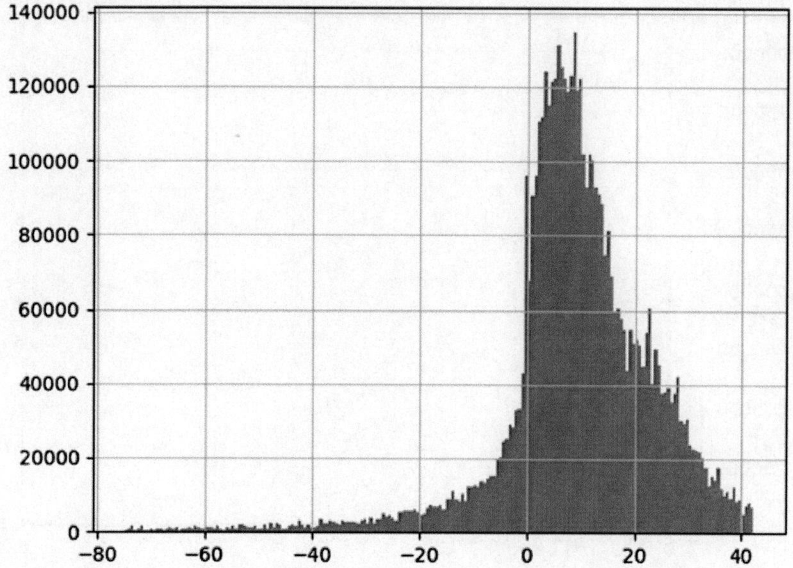

Fig. D.4 Histogram of attribute: profitmargin

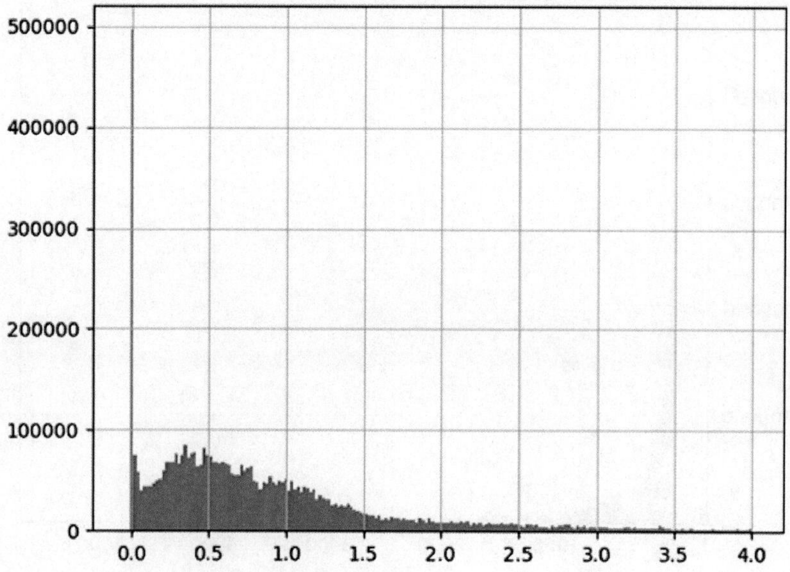

Fig. D.5 Histogram of attribute: debttoequity

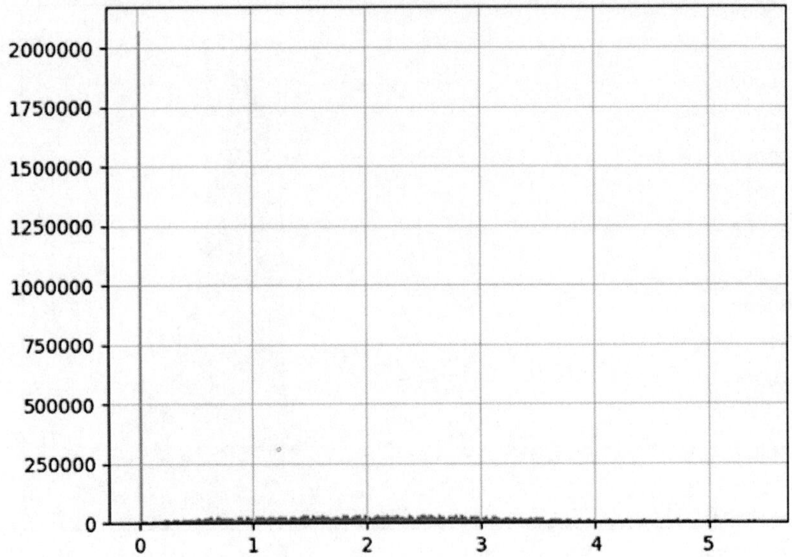

Fig. D.6 Histogram of attribute: dividendyield

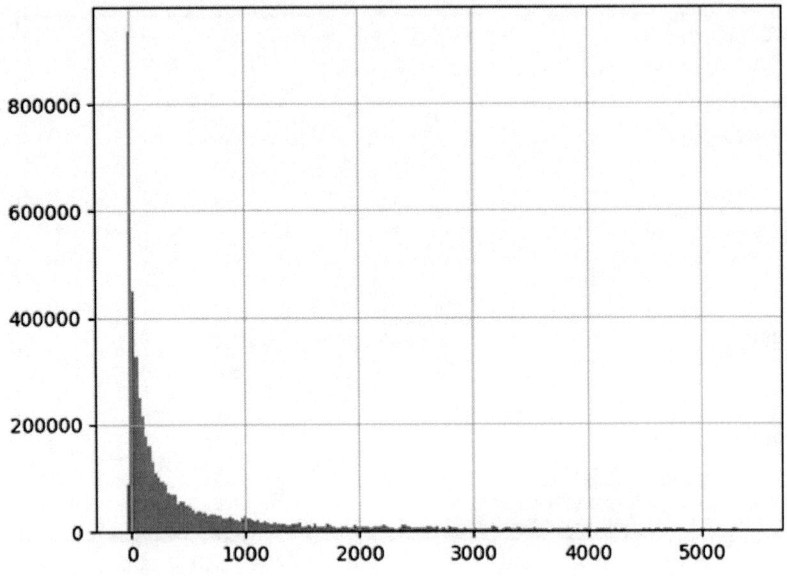

Fig. D.7 Histogram of attribute: ebitda

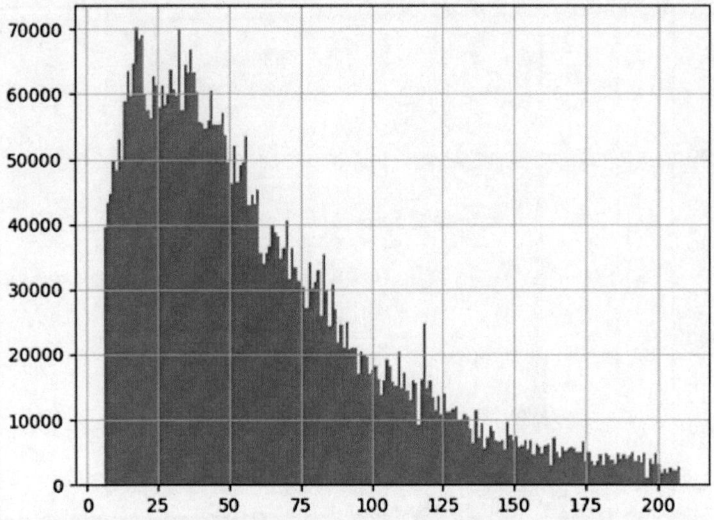

Fig. D.8 Histogram of attribute: 52weekhigh

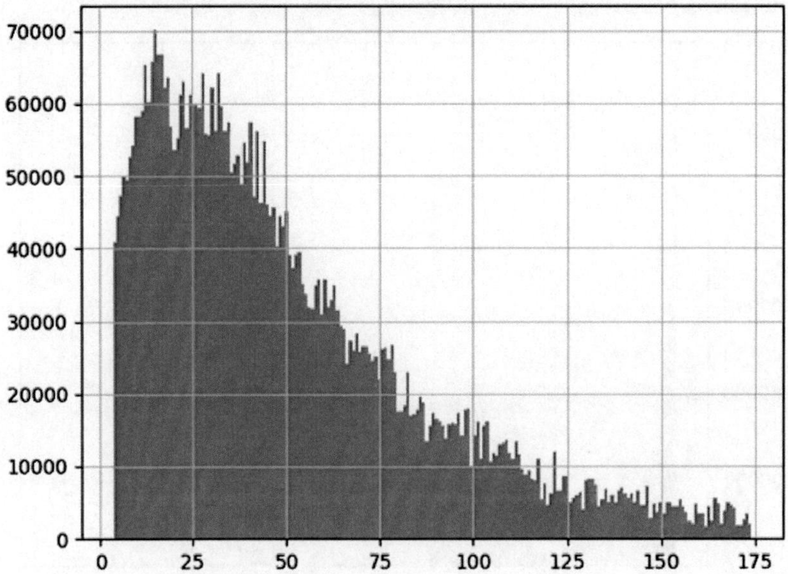

Fig. D.9 Histogram of attribute: 52weeklow

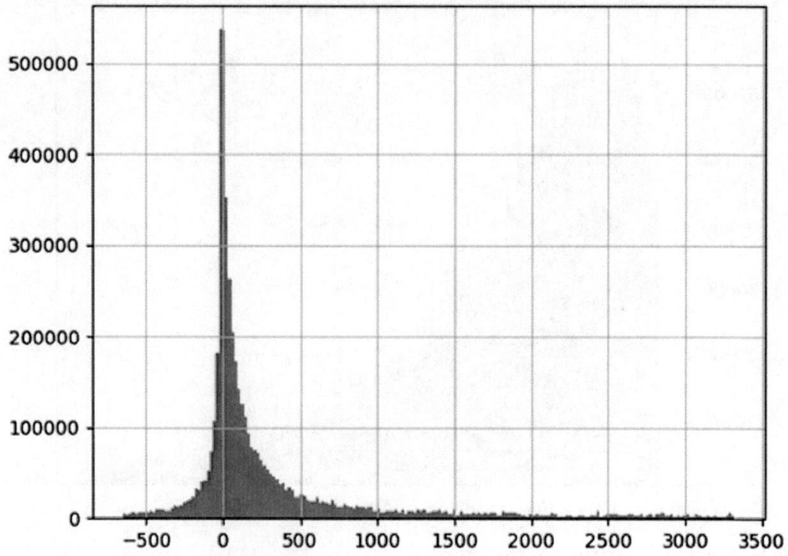

Fig. D.10 Histogram of attribute: freecashflow

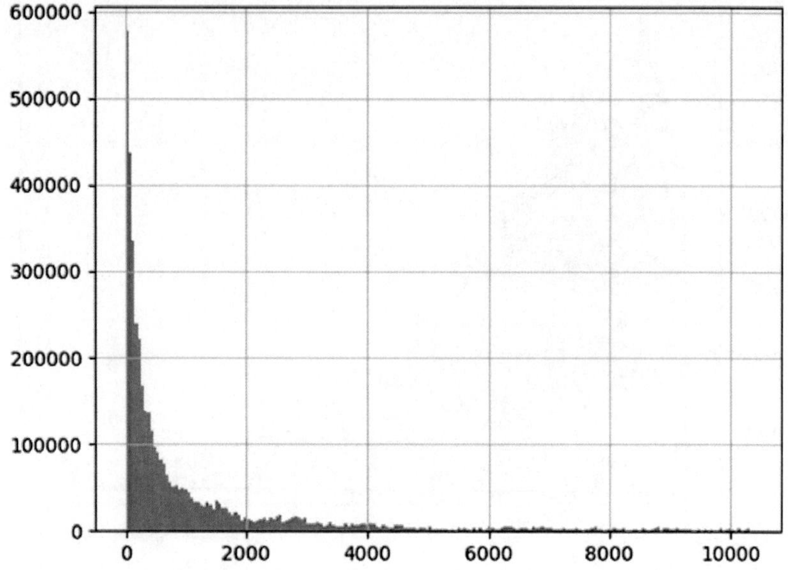

Fig. D.11 Histogram of attribute: totalgrossprofit

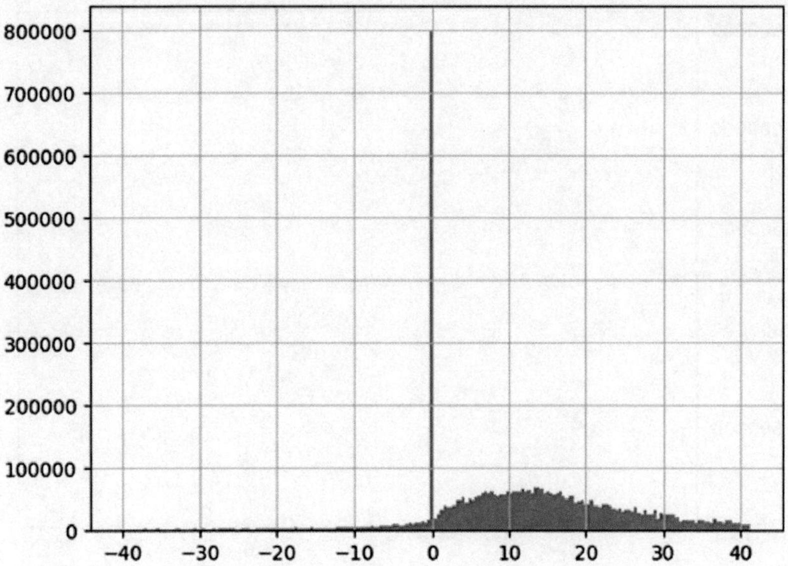

Fig. D.12 Histogram of attribute: operatingmargin

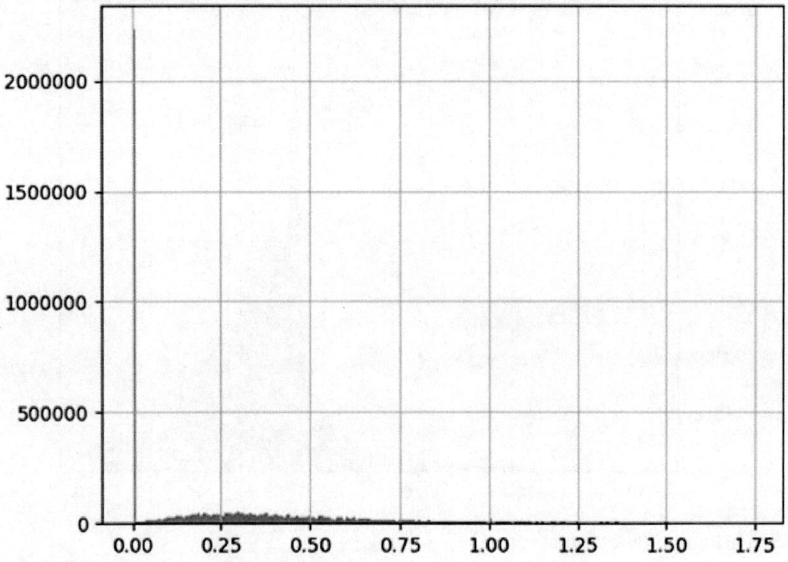

Fig. D.13 Histogram of attribute: divpayoutratio

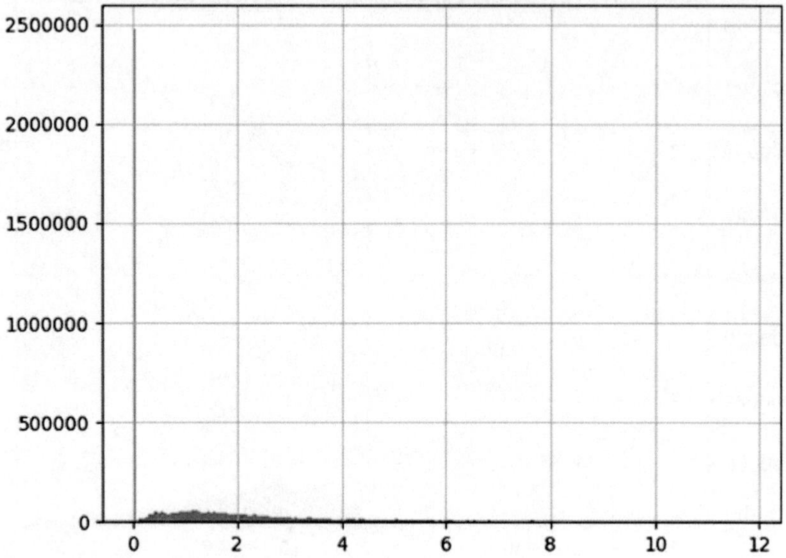

Fig. D.14 Histogram of attribute: pricetoearnings

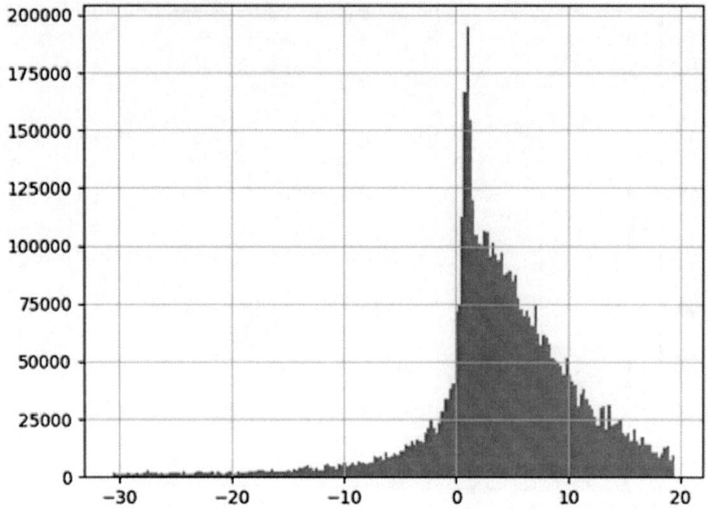

Fig. D.15 Histogram of attribute: roa

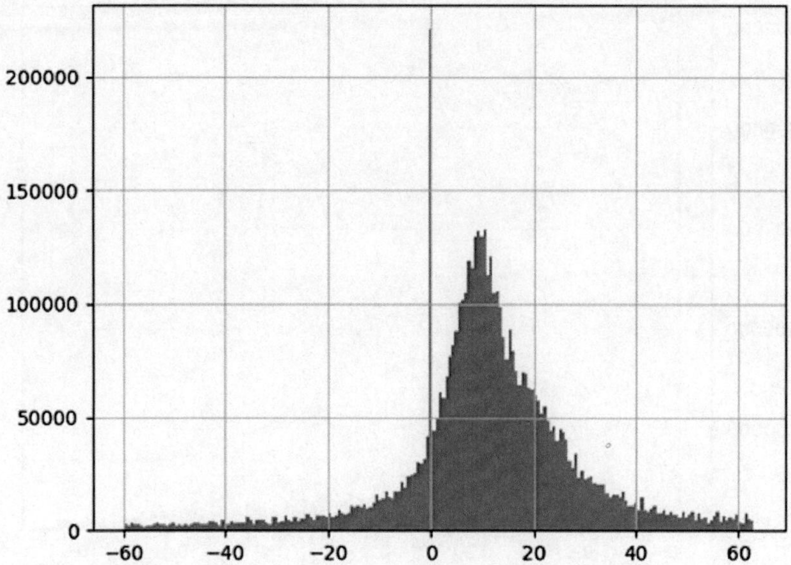

Fig. D.16 Histogram of attribute: roe

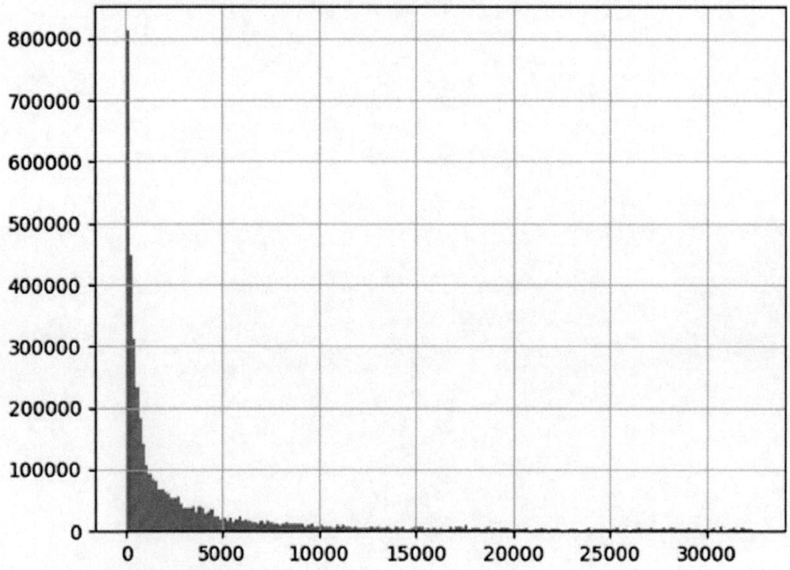

Fig. D.17 Histogram of attribute: debt

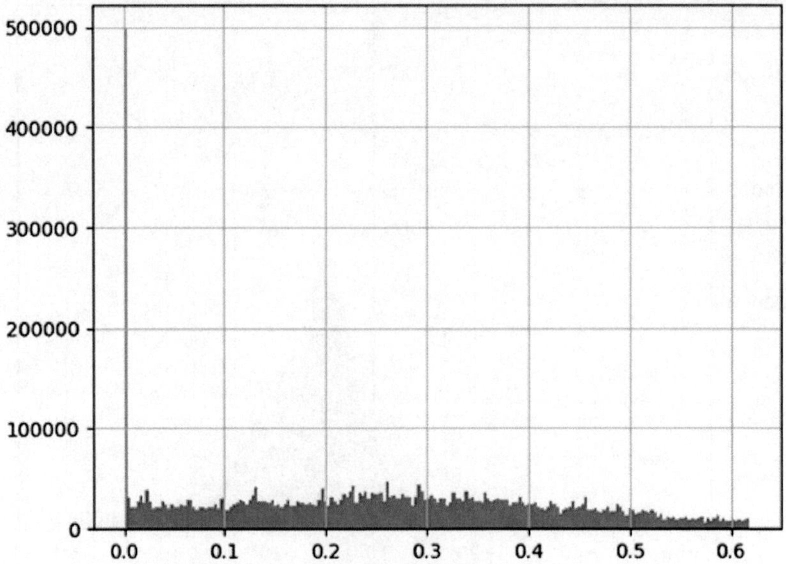

Fig. D.18 Histogram of attribute: totalDebtToCurrentAsset

Appendix E
Fuzzy Sets for Particular Attributes

Fuzzy sets determined based on the histograms presented in Appendix D are illustrated in Figs. E.1, E.2, E.3, E.4, E.5, E.6, E.7, E.8, E.9, E.10, E.11, E.12, E.13, E.14, E.15, E.16, E.17 and E.18.

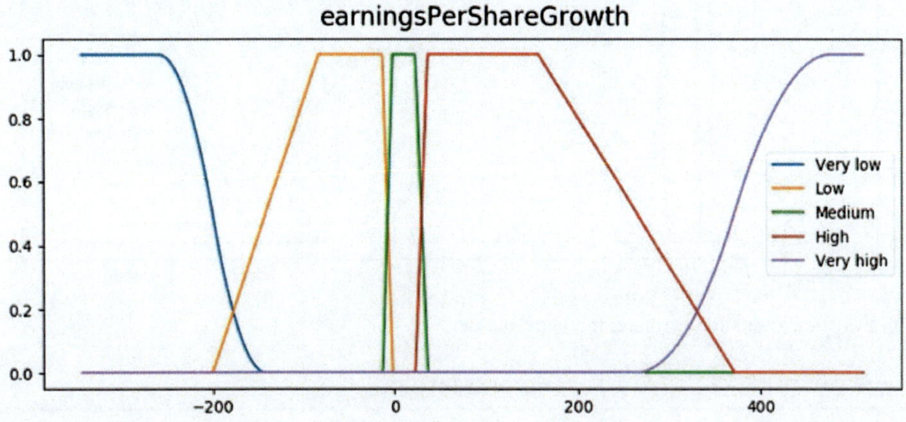

Fig. E.1 Fuzzy sets for attribute: epsgrowth

T. Rutkowski, *Explainable Artificial Intelligence Based on Neuro-Fuzzy Modeling with Applications in Finance*, Studies in Computational Intelligence 964,
https://doi.org/10.1007/978-3-030-75521-8

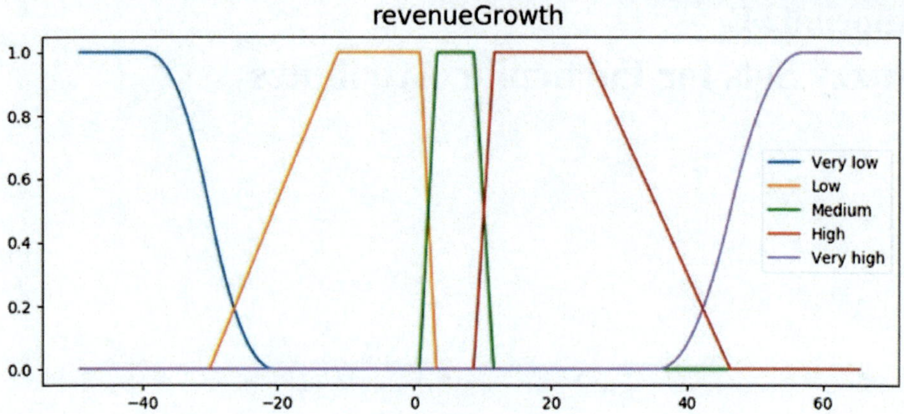

Fig. E.2 Fuzzy sets for attribute: revenuegrowth

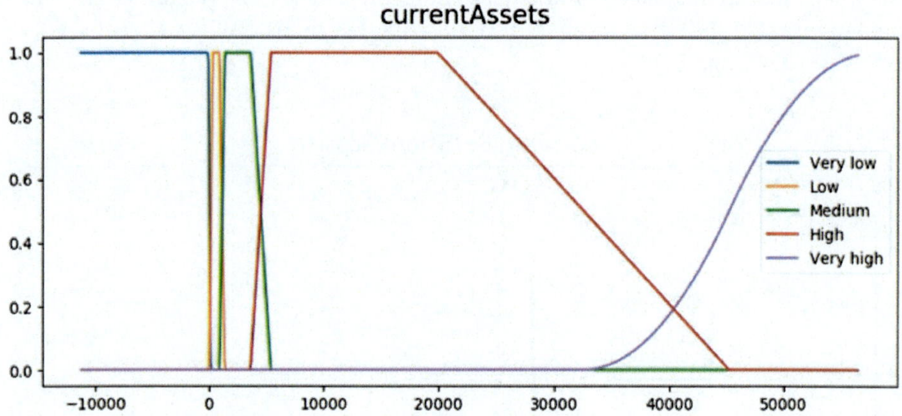

Fig. E.3 Fuzzy sets for attribute: totalcurrentassets

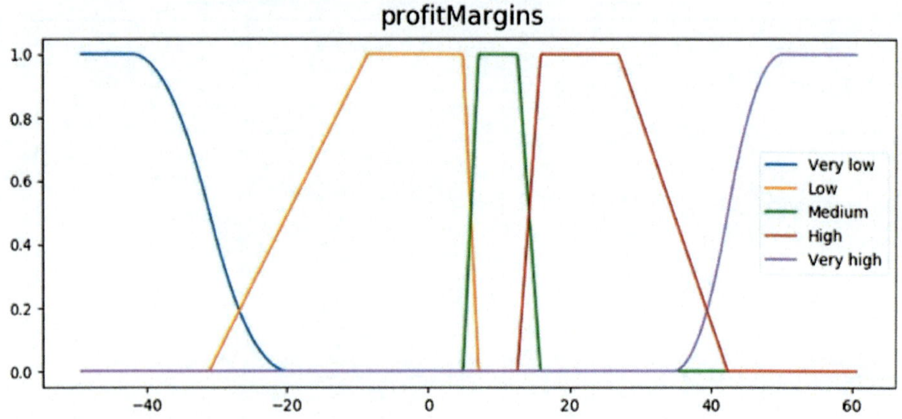

Fig. E.4 Fuzzy sets for attribute: profitmargin

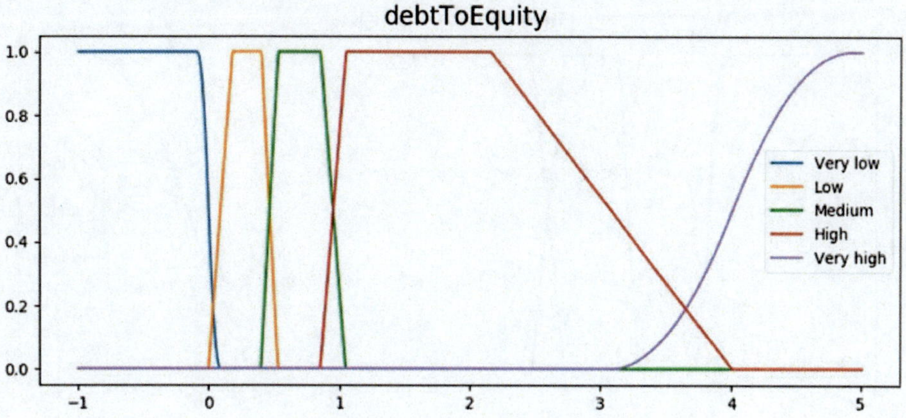

Fig. E.5 Fuzzy sets for attribute: debttoequity

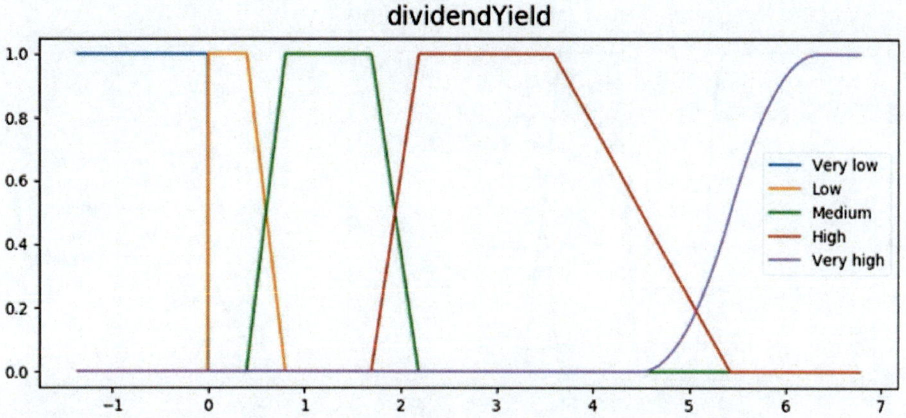

Fig. E.6 Fuzzy sets for attribute: dividendyield

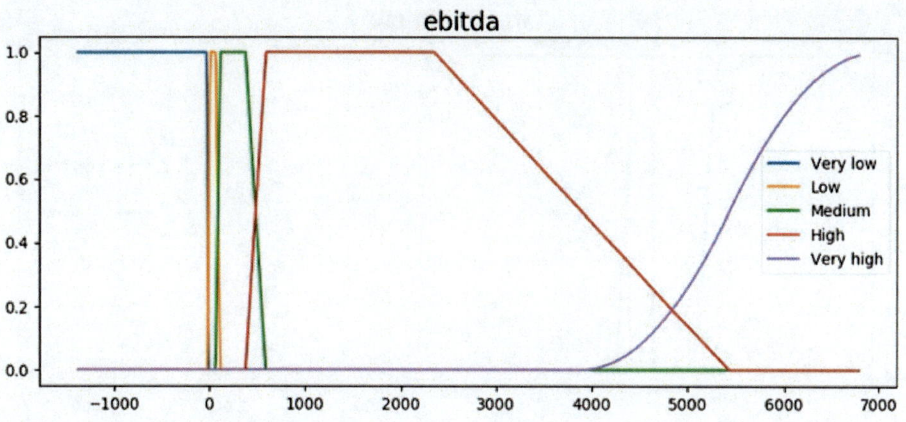

Fig. E.7 Fuzzy sets for attribute: ebitda

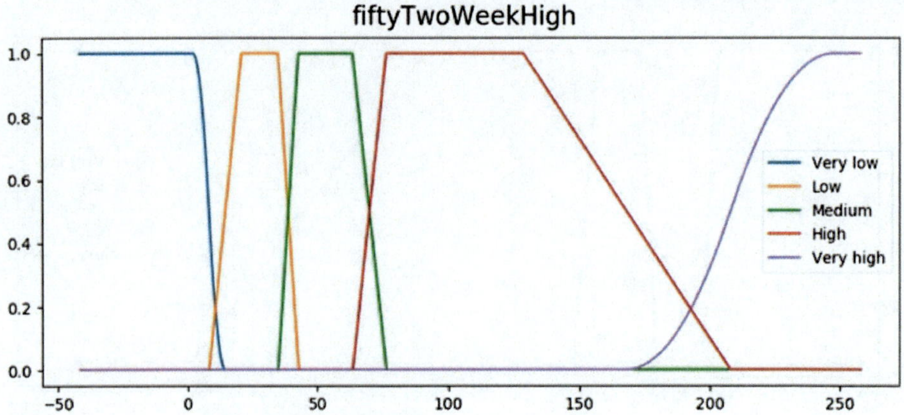

Fig. E.8 Fuzzy sets for attribute: 52weekhigh

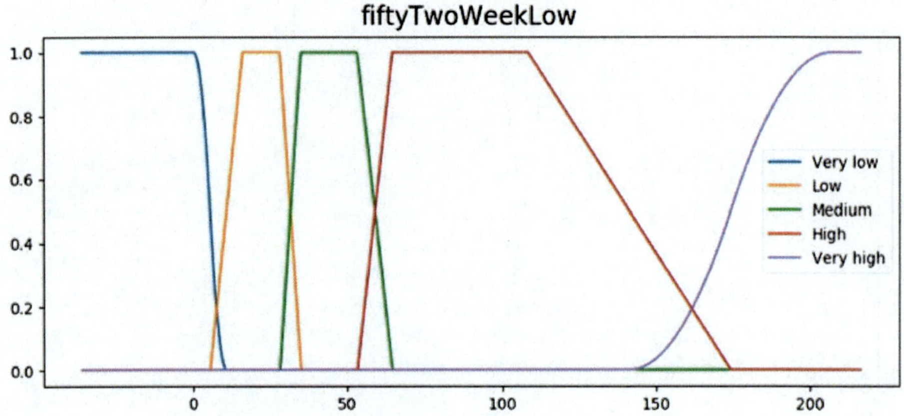

Fig. E.9 Fuzzy sets for attribute: 52weeklow

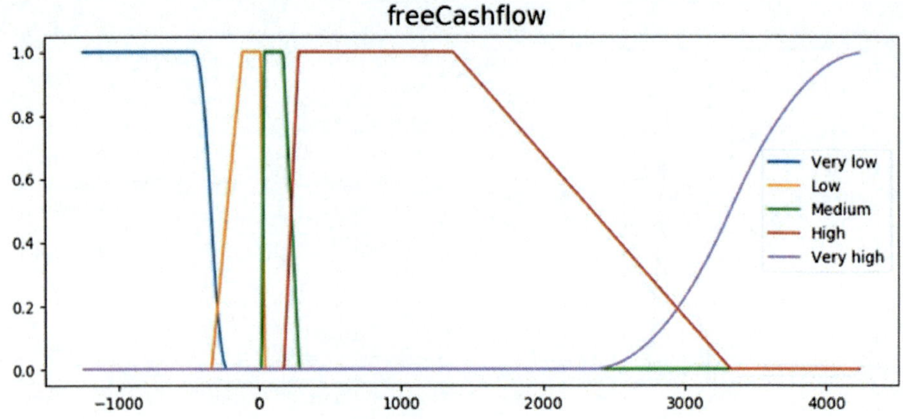

Fig. E.10 Fuzzy sets for attribute: freecashflow

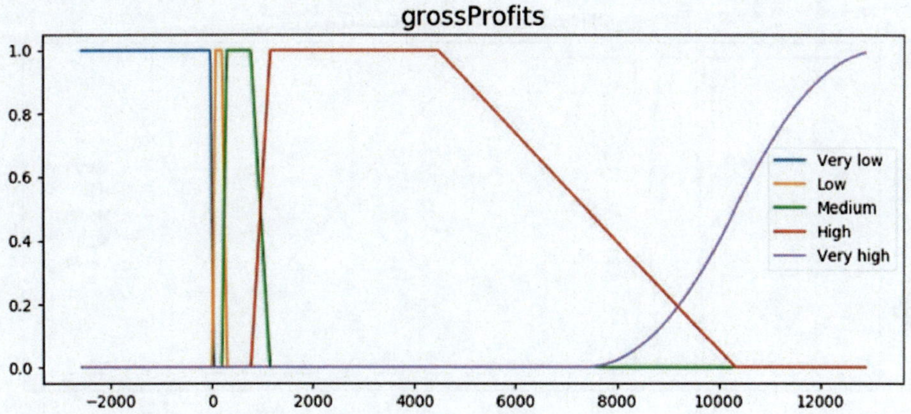

Fig. E.11 Fuzzy sets for attribute: totalgrossprofit

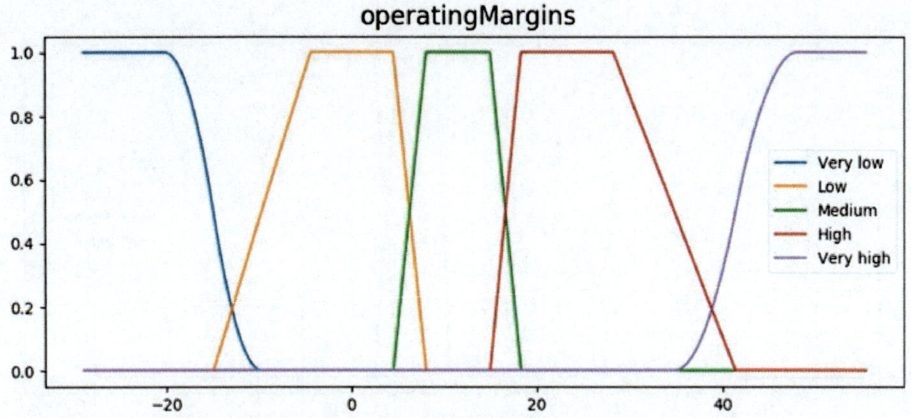

Fig. E.12 Fuzzy sets for attribute: operatingmargin

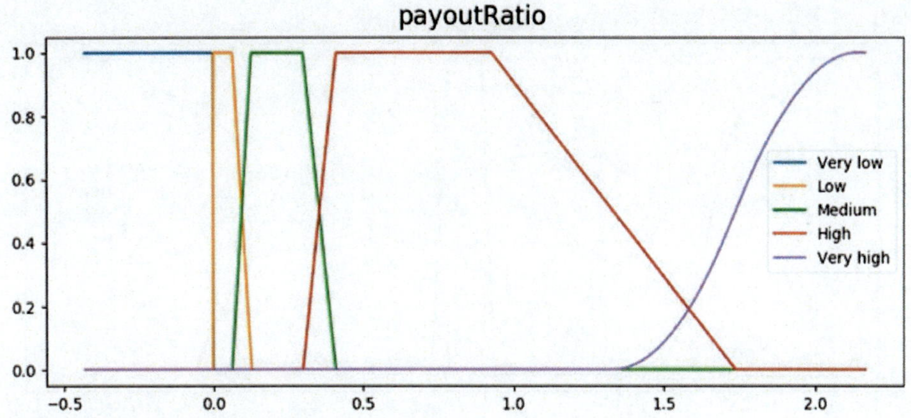

Fig. E.13 Fuzzy sets for attribute: divpayoutratio

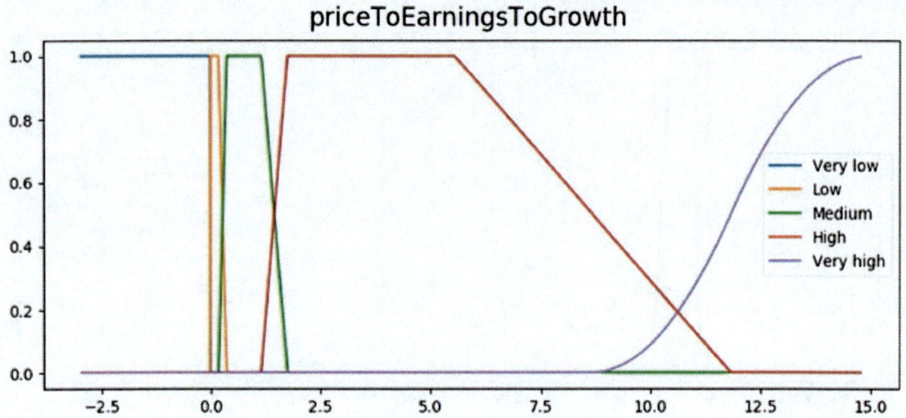

Fig. E.14 Fuzzy sets for attribute: pricetoearnings

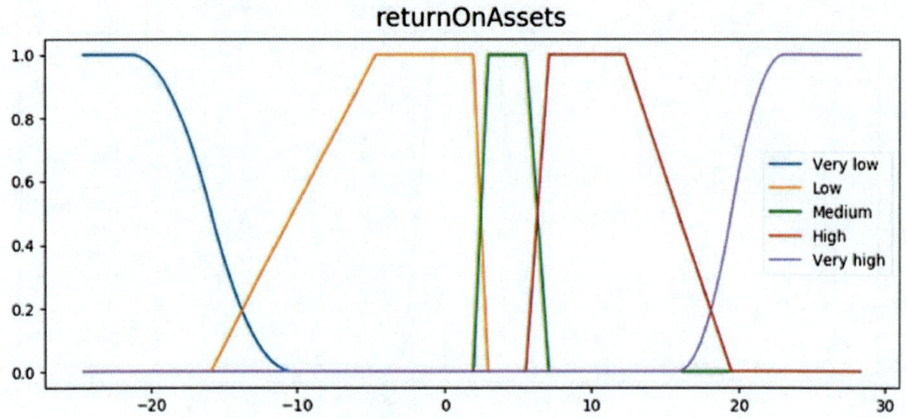

Fig. E.15 Fuzzy sets for attribute: roa

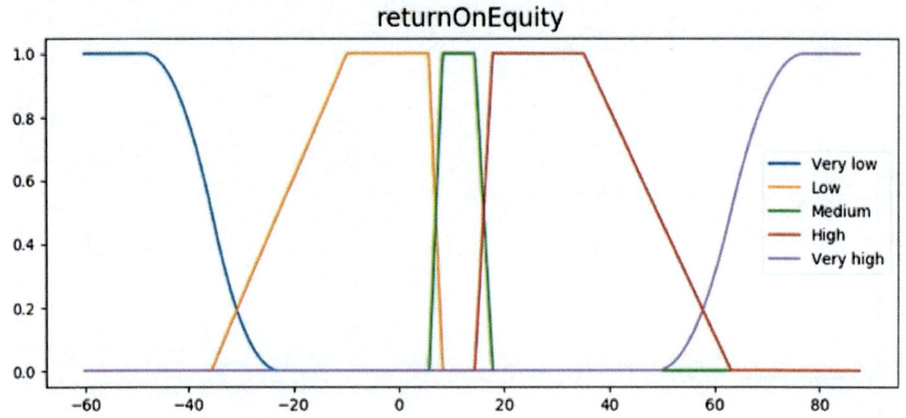

Fig. E.16 Fuzzy sets for attribute: roe

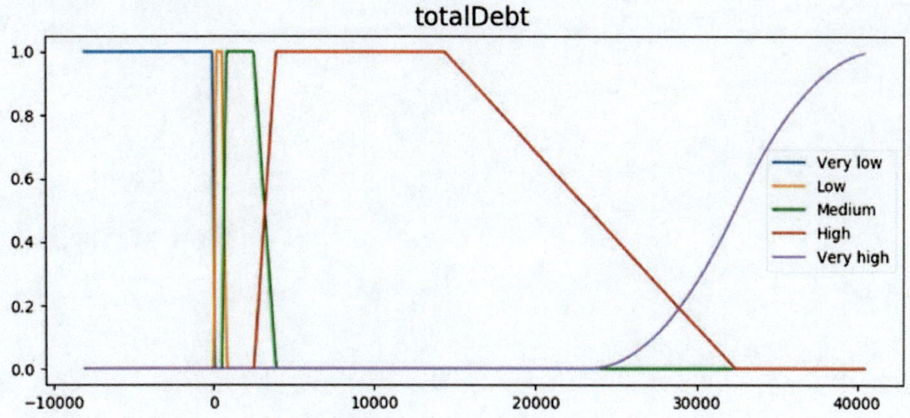

Fig. E.17 Fuzzy sets for attribute: debt

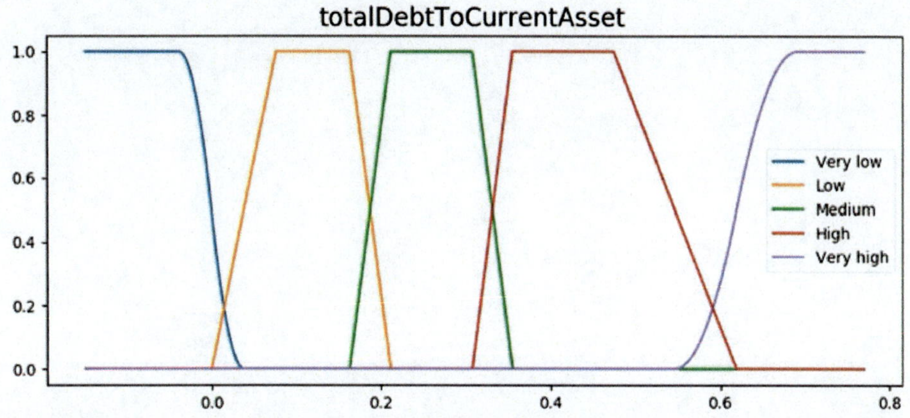

Fig. E.18 Fuzzy sets for attribute: totalDebtToCurrentAsset

Appendix F
Fuzzy Sets for Single Data Points

Fuzzy sets determined based on the past transactions of an example user (investor) are illustrated in Figs. F.1, F.2, F.3, F.4, F.5, F.6, F.7, F.8, F.9, F.10, F.11, F.12, F.13, F.14, F.15, F.16, F.17 and F.18, for particular attributes (one Gaussian fuzzy set for each past transaction).

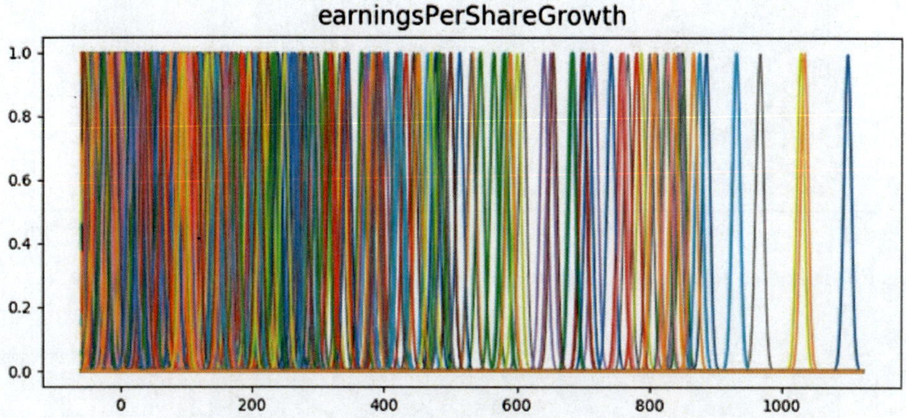

Fig. F.1 Fuzzy sets for attribute: epsgrowth—one fuzzy set for each past transaction

T. Rutkowski, *Explainable Artificial Intelligence Based on Neuro-Fuzzy Modeling with Applications in Finance*, Studies in Computational Intelligence 964,
https://doi.org/10.1007/978-3-030-75521-8

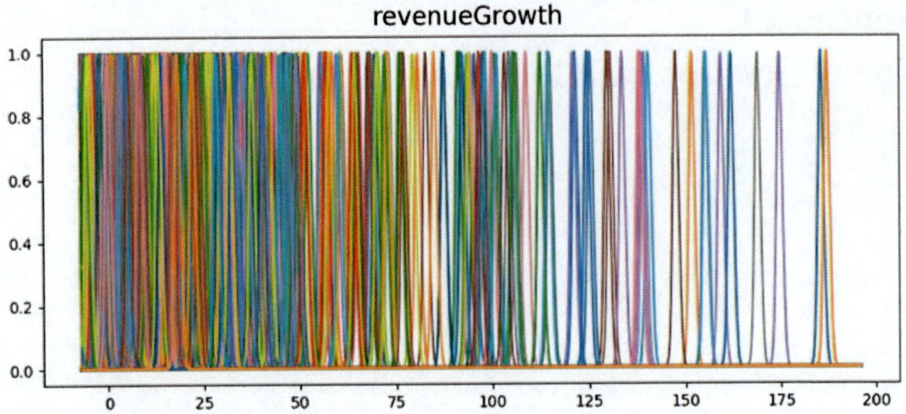

Fig. F.2 Fuzzy sets for attribute: revenuegrowth—one fuzzy set for each past transaction

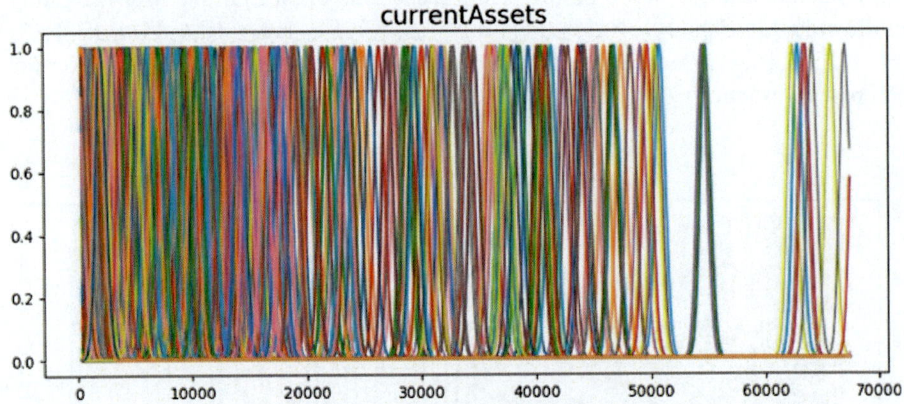

Fig. F.3 Fuzzy sets for attribute: totalcurrentassets—one fuzzy set for each past transaction

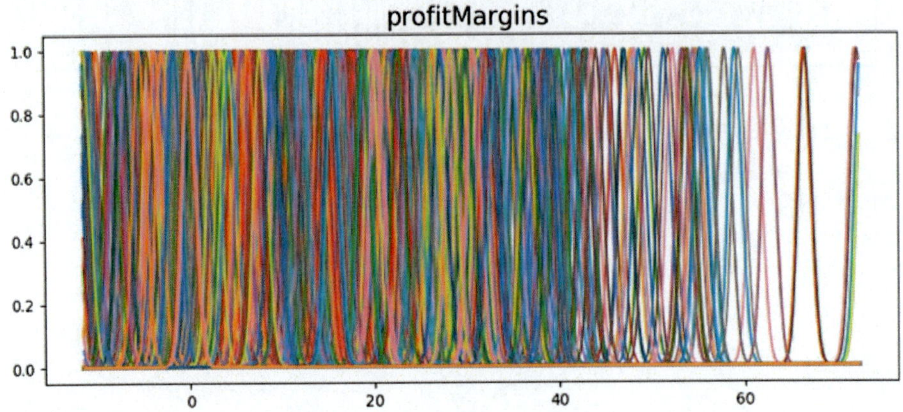

Fig. F.4 Fuzzy sets for attribute: profitmargin—one fuzzy set for each past transaction

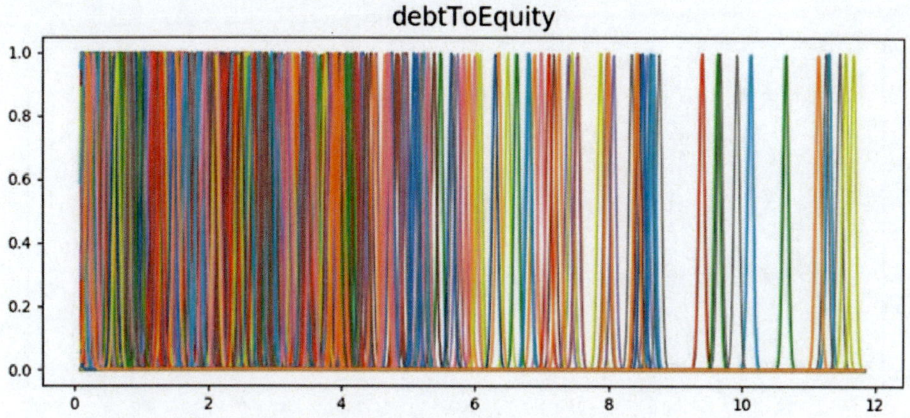

Fig. F.5 Fuzzy sets for attribute: debttoequity—one fuzzy set for each past transaction

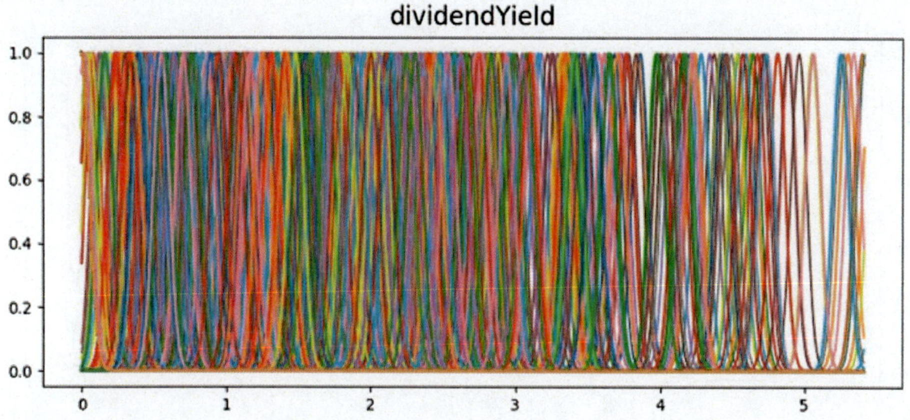

Fig. F.6 Fuzzy sets for attribute: dividendyield—one fuzzy set for each past transaction

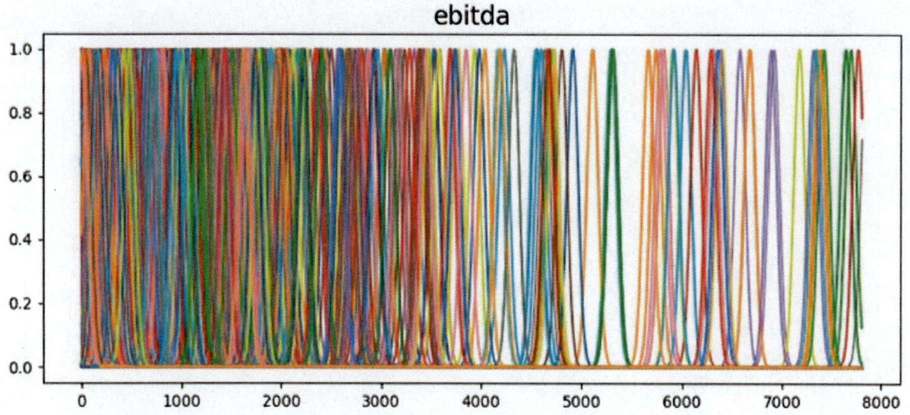

Fig. F.7 Fuzzy sets for attribute: ebitda—one fuzzy set for each past transaction

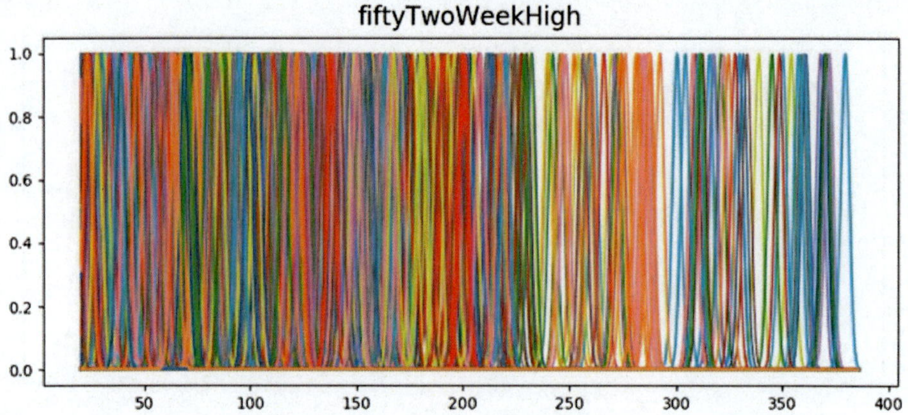

Fig. F.8 Fuzzy sets for attribute: 52weekhigh—one fuzzy set for each past transaction

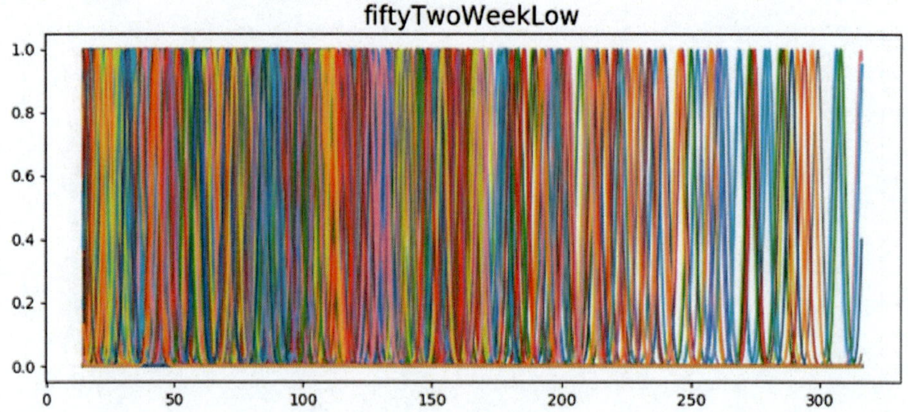

Fig. F.9 Fuzzy sets for attribute: 52weeklow—one fuzzy set for each past transaction

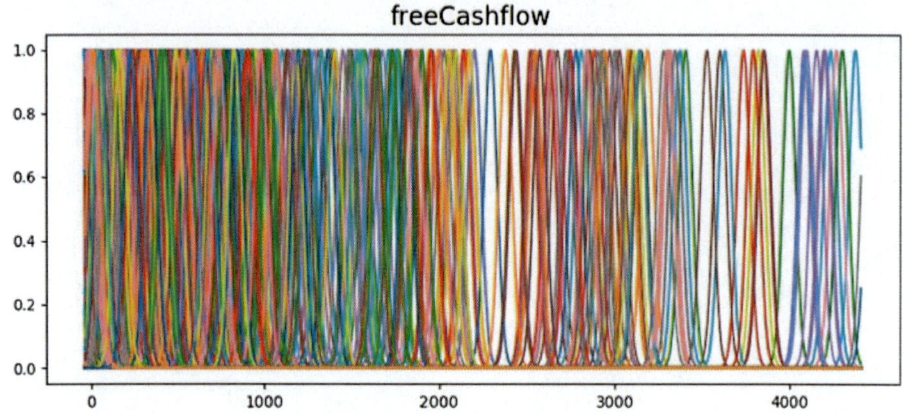

Fig. F.10 Fuzzy sets for attribute: freecashflow—one fuzzy set for each past transaction

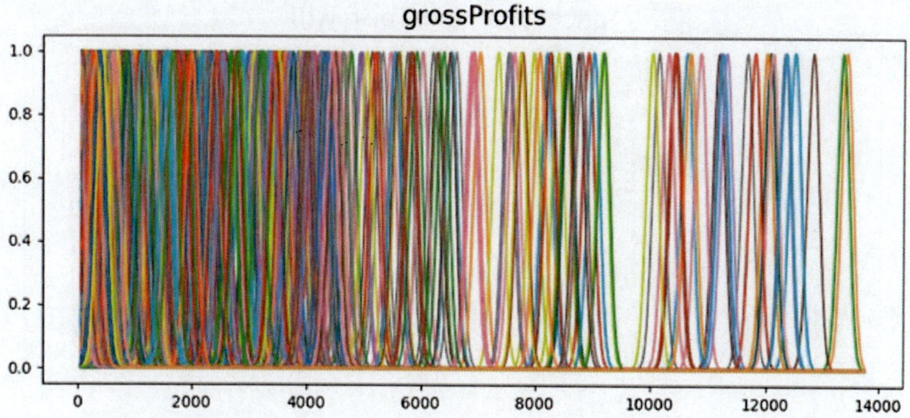

Fig. F.11 Fuzzy sets for attribute: totalgrossprofit—one fuzzy set for each past transaction

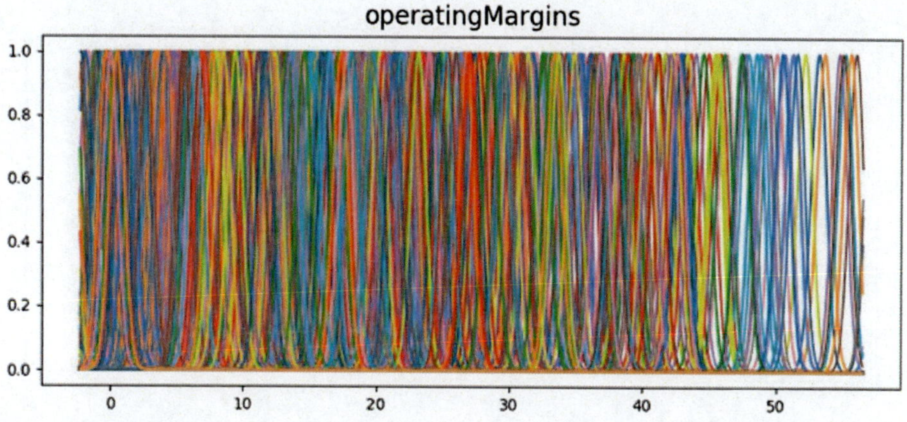

Fig. F.12 Fuzzy sets for attribute: operatingmargin—one fuzzy set for each past transaction

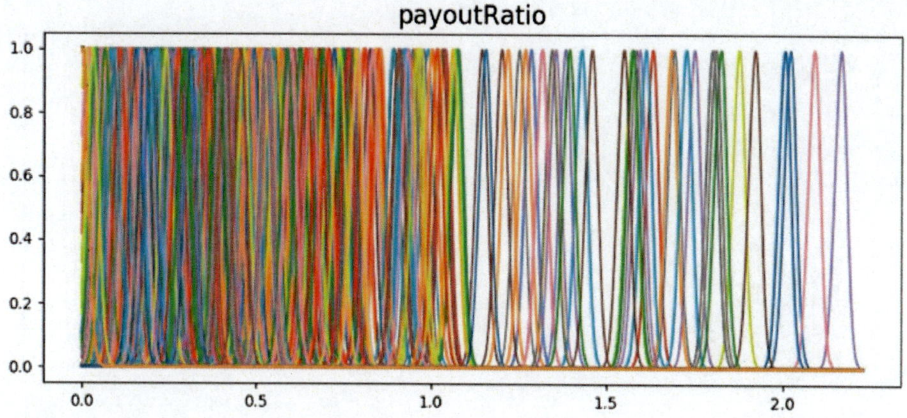

Fig. F.13 Fuzzy sets for attribute: divpayoutratio—one fuzzy set for each past transaction

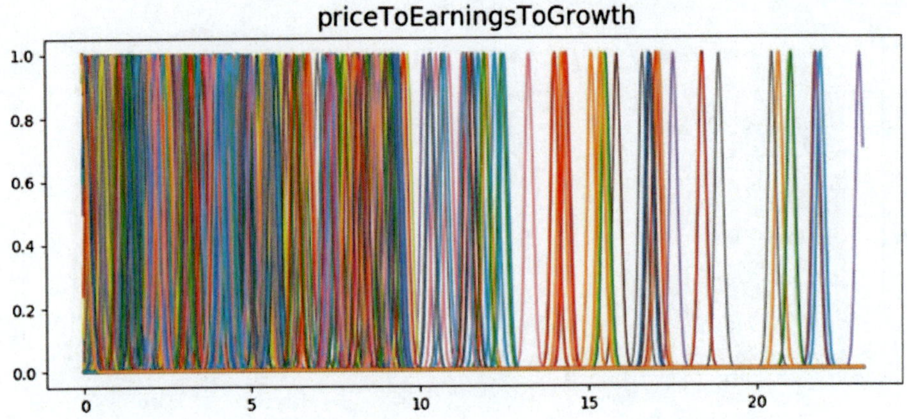

Fig. F.14 Fuzzy sets for attribute: pricetoearnings—one fuzzy set for each past transaction

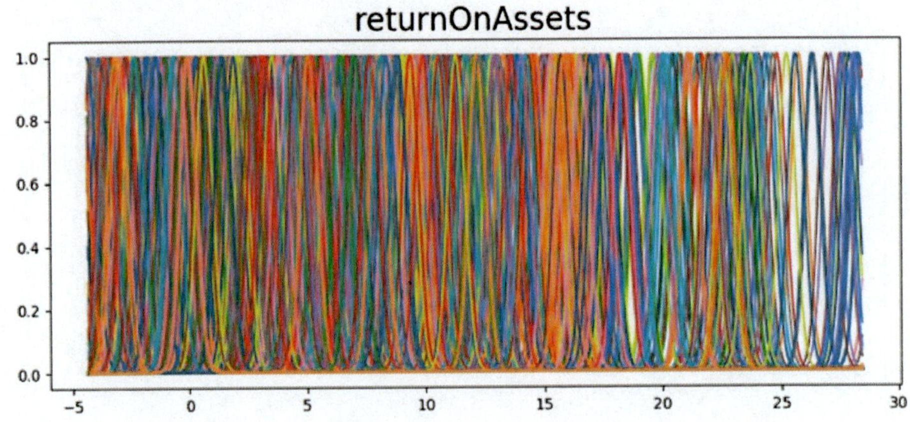

Fig. F.15 Fuzzy sets for attribute: roa—one fuzzy set for each past transaction

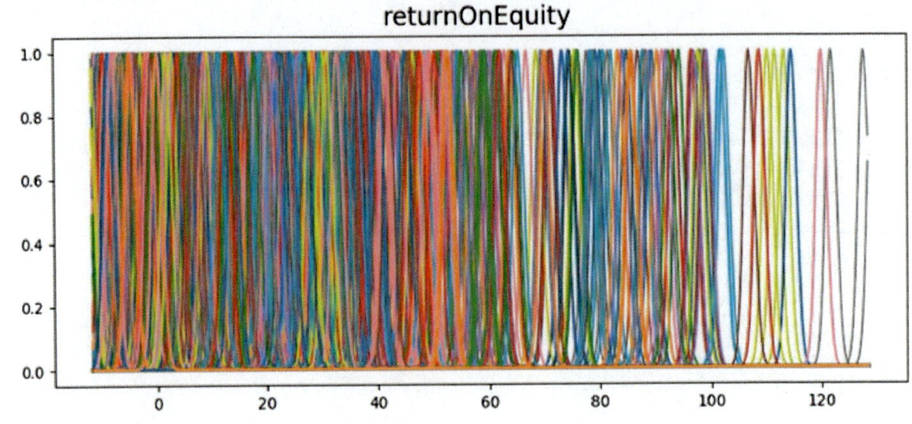

Fig. F.16 Fuzzy sets for attribute: roe—one fuzzy set for each past transaction

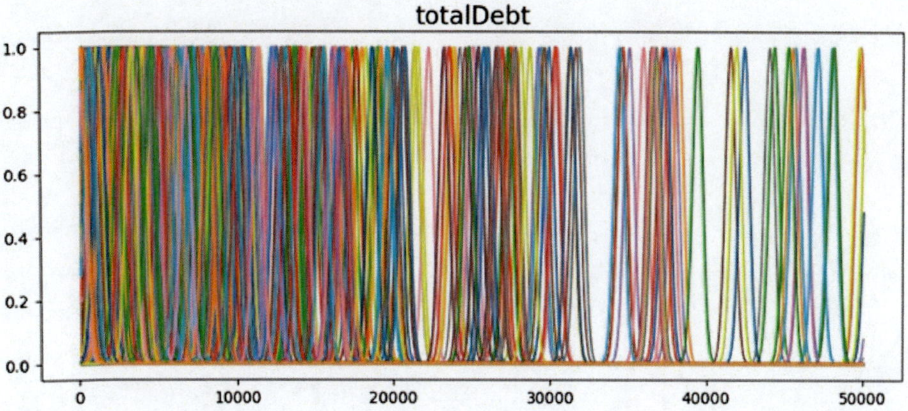

Fig. F.17 Fuzzy sets for attribute: debt—one fuzzy set for each past transaction

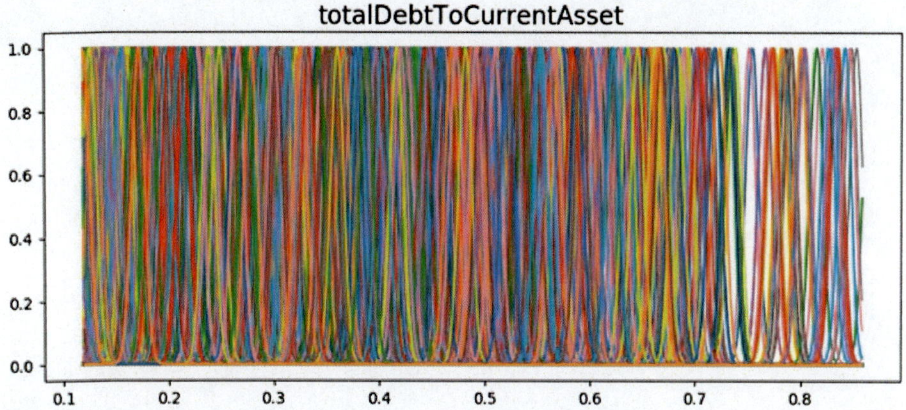

Fig. F.18 Fuzzy sets for attribute: totalDebtToCurrentAsset